BEING
SEEN

MASTER PARENTING IN THE DIGITAL AGE

SELENA BARTLETT

PHINELINE
PUBLISHING

First published in Australia in 2024 by Phineline Publishing

Email for correspondence: info@phinelinepublishing.com

© Selena Bartlett 2024

The moral rights of the author have been asserted

All rights reserved.

Except as permitted under the *Australian Copyright Act 1968* (for example, a fair dealing for the purposes of study, research, criticism or review), no part of this publication may be reproduced, stored in a retrieval system, communicated or transmitted in any form or by any means without prior written permission.

All enquiries should be directed to the publisher.

ISBN 9780999099735 (paperback)

IBSN 9780999099711 (ebook)

A catalogue record for this book is available from the National Library of Australia

Disclaimer

The views and opinions shared in the book are personal perspectives of Selena Bartlett, and based on research papers she reviewed and conversations she engaged in for the purposes of the book. The content is intended for general informational purposes for readers and does not substitute for professional medical, psychological, or healthcare advice. Although the publisher and the author have made every effort to ensure that the information in this book was correct at press time and while this publication is designed to provide accurate information in regard to the subject matter covered, the publisher and the author assume no responsibility for errors, inaccuracies, omissions, or any other inconsistencies herein, and hereby disclaim any liability to any party for any loss, damage, or disruption caused by errors or omissions. These opinions do not represent the views of Queensland University of Technology.

For the children we need to see

Contents

Foreword ... vii
Preface ... 1
Introduction ... 7

PART 1 THE PILLARS OF PARENTING

1. Being Seen ... 17
   Understanding its importance in brain and child development
2. Seeing Yourself as a Parent ... 37
   Cultivating self-awareness
3. Know Your Actions ... 53
   By seeing your child's silent imitations
4. Feeling Connected ... 67
   To yourself, your child and others
5. Being Tech and Sex Savvy ... 75
   Having digital literacy skills

PART 2 BARRIERS TO PARENTING

6. Living in the 21st Century ... 89
   Parenting children in a stressful world
7. Echoes from Our Past ... 99
   How our childhood shapes our parenting
8. The Technology Factors ... 113
   The serious challenges for parents in the digital age

9.	Lost Connections *From yourself, your child and others*	127
10.	Lifestyle Factors *The unseen impact of poor diet and lack of exercise on parenting*	137

PART 3   STEPS TO MASTER PARENTING IN THE DIGITAL AGE

11.	Managing Stress *Rewire your overwhelmed brain*	151
12.	Re-parenting *Invaluable benefits for children and future generations*	169
13.	Unlock Tech-Savvy Success *Family tech plan, tech-free zones, social media and screen time*	187
14.	Create a Place Where Everyone Feels They Belong! *Build a village where you are being seen*	215
15.	Living a Vital Life *The power of whole foods and fitness*	231
16.	Take Forward Action *Moving and playing*	241
17.	Being Sleep Wise *Sweet dreams*	253
18.	Prioritise Your Brain Health *Never lose sight of what really matters*	263
Conclusion		271
Acknowledgements		275
About the Author		277

Foreword

Over the years, I've balanced the roles of community builder, entrepreneur, and parent to three amazing children. The experiences have taught me invaluable lessons about the evolving challenges that parents grapple with, especially in this rapidly changing digital landscape. That's precisely why ParentTV came into existence.

In today's digital landscape, parents are encountering hurdles that have never been faced by any generation of parents in the past. My resolve to establish ParentTV stemmed from a year that shook my community to its core—marked by the loss of young lives, including my son's 19-year-old kindergarten teacher and a 14-year-old former student of my husband's, both dying by suicide. Their untimely departures, along with others', signalled an alarming truth: an increasing disconnect in our society, escalating mental health issues among our young people, and a clear need for change.

With disheartening statistics like one in ten children entering kindergarten showing symptoms of mental health conditions, my commitment to acting as an agent of change intensifies daily. This book, *Being Seen*, written by a distinguished neuroscientist and fellow parent, Selena Bartlett, serves as a much-needed compass in these turbulent times. Crafted out of both scientific rigour and a mother's love, this book encapsulates the author's dedication to enriching the lives of families everywhere. The chapters aren't just a compilation of advice; they're more like a heartfelt conversation with you. One essential theme this book tackles is the notion of "re-parenting ourselves". In essence, self-awareness and self-care lay the foundation for more effective parenting.

This becomes even more relevant when it comes to helping our children navigate the digital universe. Understanding the intricacies of their online interactions requires a certain level of personal grounding. Therefore, *Being Seen* offers practical tools for engaging in essential, open conversations with our kids about their digital lives. A central aspect of this book is the idea of the "serve and return" relationship—a focused, emotional exchange with our children that fosters their emotional well-being and resilience. This powerful concept is often missing in traditional parenting advice, yet it's a cornerstone for cultivating lifelong health. Additionally, the book shines a light on an often-neglected aspect of family well-being: brain health. For both children and adults, understanding how our brains work can significantly influence our ability to parent effectively.

Our collective mission is to create a world where our children are not just seen but are truly nurtured—where technology is our servant and not our master. We are committed to building stronger, more connected parents, children and families and tackling the pressing challenges of our time. By choosing to read *Being Seen*, you're joining us on a collective mission: to create a world where children are seen, heard, and nurtured—a world where technology augments our lives rather than dominating them. Together, we will light the way to a future where the essence of parenting and technology harmoniously coexist, focused on what truly counts.

This book is your roadmap to making that future a tangible reality for us all.

Let's do good things together!

Sam Jockel
Founder and CEO of ParentTV and Producer of SEEN Documentary

Preface

As parents, we are always doing the best we can; we love our children. We want the best for them. However, parenthood and caring for children in the digital age is almost impossible. As both a neuroscientist and a parent, I wrote this book with you in mind. It's the guide *I* needed after welcoming my first child into the world. There's the real struggle and the juggling act between work commitments and family responsibilities; the financial pressure, guilt, stress, or anxiety about not spending *enough* time with our children, or the fear of missing important moments in their lives. There are the sacrifices we make by putting certain aspects of our lives or careers on hold for the sake of our children. There's the continual worry or fear as to whether we are doing the right thing when it comes to how we raise our children. Then, there's this new technology that creates an invisible hurdle that no one quite knows how to navigate. It's safe to say parents and carers face exhausting circumstances, often with little or no support. As you read the chapters of this book, think of it as a heart-to-heart conversation, an exploration of possibilities and hope. Taking the time to read and dive into this topic shows just how much you care about your children and their children.

This book is the culmination of three decades of my journey as a neuroscientist, sparked by a personal quest to understand and support my sister, Francesca, in her battles with mental illness. While this was the initial motivation, this expedition transformed and enriched my perspectives not just as a scientist, but also as a parent and an individual. At its core, ***Being Seen*** is about truly recognising and valuing someone for who they are. ***Being Seen*** distils the complexity of human experiences, feelings and emotions and translates it into one simple action that can help everyone thrive. We can genuinely notice and understand someone's journey, challenges, and feelings when we are being seen ourselves. This is the secret formula that leads to happy and thriving people, no matter where they come from or what they've been through. It's something so simple, yet so powerful.

Francesca's challenges with mental illness were seen in the wrong way, and the ensuing responses from medical professionals, our community, and within our family were veiled in a hushed tone of secrecy and embarrassment. As time went on and my knowledge expanded, I realised a profound truth: recognising and prioritising brain health is one of the greatest gifts we can give ourselves and our children. This insight not only paves the way for success but can also prevent or more effectively tackle mental health issues. Embracing brain science reshaped how I viewed myself, influencing my relationships and parenting approach. I began to truly *see* myself and my children.

So, as caregivers, where do we begin? The key is in seeing ourselves in a new way, recognising the wonders of the brain and its remarkable ability to adapt, known as neuroplasticity. Central to understanding brain health are the intersections of our early life experiences and genetics, both of which leave profound imprints on our brain development and overall mental well-being. Think of our brains as the master conductors, orchestrating everything from our deepest feelings to our most intricate actions. These incredible hubs of activity, while mighty, require nurturing and care to operate optimally. That's the essence of seeing that brain health is everyone's business and not only someone else's problem.

PREFACE

Undeniably, the digital age has made parenting even more difficult. I watch helplessly as people are barely able to keep up with the technological advances or are racing to secure smartphones, apps, games, devices—unaware of the dangers that lurk behind the screen—or perhaps knowing but unsure how to handle it all. And this creates relentless pressure on both parents and children. Teachers are equally strained. Schools are struggling to handle the mental health and well-being of their students. And it seems like everyone, from governments to corporations, is more worried about data protection than our children's well-being. Unfortunately, this is the reality that many parents are facing, and it shows we're in a battle we didn't sign up for. Because of technology, parenting has become more complicated compared to 10 or 20 years ago. We don't only have to look out for our children in the real world; we also must keep them safe online.

According to the US Surgeon General's report *Social Media and Youth Mental Health*, a staggering 95% of teens between the ages of 13 and 17 are active on social media platforms. Even more eye-opening is the fact that nearly 40% of children aged 8 to 12 are also engaging with these platforms. This rising trend raises significant concerns among parents, caregivers, healthcare professionals, and researchers about the potential impact on the mental well-being of our young people. Complicating matters further is society's own pervasive addiction to smartphones and social media.

As devices, apps and technology become more ingrained in our everyday life, preventing our children's access to them is a losing battle. I've encountered many stories in Facebook groups dedicated to parenting in the tech world, and some of them are heartbreaking. Many parents feel lost and all they can do is share their stories and ask for help from other parents.

The stark reality is we, as parents, are stuck in a digital age that we must navigate. The heart of this book is about helping parents navigate this world and learn a new digital language and parenting skill set that we could not have been taught by our own parents—one that is fit for purpose in the digital age. One of the primary protective factors for children's brain health and safety is *being seen* by their parents and caregivers through quality time and interactions, also called "serve and return" relationships. This is rarely discussed in parenting or pregnancy classes. This type of relationship means you are *seeing your child* rather than a device or a screen, meaning *you*, as a parent or caregiver, offer your devoted time and attention in a way that makes them feel safe, secure, and connected. If you learn to do this, you will understand your child in a new way, and the strength of the connection is a primary way to ensure their lifelong well-being, security and safety.

In today's digital landscape, it's crucial to ask: Who is seeing and interacting with your children online? And what are they asking from them? Understanding the signs of online grooming and exploitation is now a non-negotiable skill for 21st-century parents. This book aims to equip you with the knowledge, the vocabulary, and a set of questions to open crucial dialogues with your children. It is possible to learn how to conduct open-ended conversations, learn a digital language, and build the confidence you need for parenting children in the digital age. It means learning how to become comfortable with difficult conversations and talk to children about tech and sex in an age-appropriate manner. It's opening our eyes as wide as possible and paying close attention to what our children are doing rather than an idealised version of what we hope they do.

Since, as parents, we are often time poor, this book narrowly focuses on only a few key areas for parenting children in the digital age. The goal isn't to add stress to an already stressful situation. Rather, the hope is to help you understand the importance of brain health in nurturing thriving minds (for yourself and your child). In the book, you'll find straightforward insights from neuroscience into how the brain develops and is influenced by the environment, early life experiences and social connections. Throughout this book, I use the word "parent" to reflect any adult in the care of a child, to make the reading easier and more straightforward. Because children primarily learn from their parents, carers, educators, and coaches, my aim with this book is to help raise awareness of the importance of our own behaviours, acknowledge them, and put effort towards changing them. It is never too late to become the people we've always wished to be and, thus, positively influence our young and adult children's lives and development.

The book is organised into three parts. The first part discusses the pillars of parenting, from building strong connections to fostering an environment of open communication. We'll explore what truly matters in raising resilient, confident, and compassionate individuals. The second part identifies and confronts the barriers to parenting, helping us understand, acknowledge and navigate them. The third and final part presents an in-depth guide for how to master parenting in the digital age, with strategies, insights, and reflective exercises to equip parents for the challenges of raising children in today's interconnected world.

So, let's take this journey together. Let's envision a world where we are being seen—by healthy, attentive, and loving people—and where we are seeing that our children are thriving. A world where technology is our tool, not our master. Let's light the way for a brighter future—navigating our complex world and never losing sight of what truly matters. Ultimately, this book is your guide to making that world a reality for all of us.

Selena Bartlett
November 2023

Introduction

The well-worn expression "I'm at my wits' end" resonates deeply for many parents navigating the digital age. One sentiment I often hear is, "The more I try to regulate my 12-year-old daughter's time online, the cleverer she gets at evading my rules." The challenge of parenting in this digital landscape is proving to be both real and complex. "I found secret Snapchat and Pinterest accounts she set up to sidestep the restrictions we had set, so I blocked Pinterest and Snapchat entirely. And now, while she's supposedly doing maths homework, I see her activity online filled with Google searches for answers." Again, it feels like a losing battle, where containment only leads to more spillage. At this point, the idea of moving to a remote farm devoid of wi-fi and screens is becoming increasingly appealing.

The situation has evolved beyond what I'd imagined it could. Now, kids are diving headfirst into a digitally enabled and immersive world where parents are struggling to keep up. As you probably already know, most parents are time poor and mostly do not have the digital literacy or skills of a well-staffed IT security department to support them. Furthermore, we live in a society where we don't have the language skills to talk about tech or sex with our children in an age-appropriate manner. Sex as a topic of conversation is often frowned upon and dismissed—and, generally, avoided. I've heard stories of parents receiving notifications that their child has accessed a new messaging app they have never heard of, after they just banned their child from talking to a certain person. Big companies spend loads of money to keep their data safe ... but what about our children?

Most of us must figure it out on our own, relying on Facebook groups or advice from friends to make our kids' online experience safer. But what if you find out your child has been dodging screen-time limits by changing the time zone on their phone? You'd feel overwhelmed, right? Parents, we have the power to make changes because we're the ones who care the most for our children and we vote. It's time to draw a line in the sand. No more. We must demand stricter regulations on technology for children—at least for those under 13 years of age. In the meantime, we cannot wait for regulations. We must act now, and we can do this together and find immediate ways to keep our children safe and promote their mental health and well-being. You can learn skills that support parenting and make the journey a little easier.

I remember all too well the night before my son was born. I was in my university neuroscience lab and focused on finishing experiments, expecting to be back at it within just a matter of weeks after giving birth. Little did I know, my life was about to change forever (how could it not?). The day I left the hospital with my beautiful boy, I had absolutely no idea what I was doing—as I'm sure many parents and carers will understand. We aren't given any manual or guide on how to raise our kids. Most of what we learn is from our own parents. This is usually our innate parenting or carer framework, often setting the stage as to how we will behave as parents ourselves. The only manuals I had were about breastfeeding and how my son should be positioned while he slept. There was no guidance on how to help steer my children towards becoming successful adults, especially during those initial and impressionable first years.

With a PhD in brain science, I thought I knew everything about the brain and how it worked. But nothing could prepare me for the complexities of parenting. Under the stress and pressure of being a new parent, I found myself relying on the only parenting style I knew: the way my own parents raised me. But I quickly realised that this wasn't enough. I needed new skills and ideas for raising a child in the modern world. And it wasn't until years later, with plenty of trial and error, that I discovered how important early life experiences truly are in shaping a child's brain development and mental health.

When we think about our children growing up, we often focus on their physical growth—their first steps, first words and even their first day of school. But beneath the surface, there's another kind of growth happening, one that's just as crucial: their brain development. Have you ever wondered how certain childhood experiences, especially challenging ones, can shape the way our children think, feel, and behave? Research has uncovered that early life experiences, especially those tough times known as adverse childhood experiences (ACEs), play a significant role in shaping the mental and physical health of a child's brain across the lifespan.

It might sound a bit alarming, but here's the good news: amidst all the challenges, there's a powerful antidote—our relationships with our children. The love, care, and connection we give and share with our children is like a protective shield, guarding their brain from negative influences and paving the way for resilience. These special bonds between parents and children do wonders for their brain, impacting it in three main ways.

Firstly, during the early years of our children's lives, the brain is developing at an incredible pace, with neural connections being formed at a rapid rate. The *quality* of our interactions with our children can have a significant impact on the formation of these neural connections. Positive interactions, such as affectionate touch, responsive communication, and sensitive caregiving, can promote the development of neural networks involved in emotion regulation, social cognition, and stress response. On the other hand, negative interactions, such as neglect, abuse, and harsh parenting, and now tech exposure too early, can *interfere* with the formation of these neural connections and lead to the development of mental health problems later in life.

Secondly, the relationship between parent and child can shape a child's beliefs and attitudes about themselves, others, and the world around them. For example, a secure attachment relationship between a parent and child can promote the child's sense of security, trust, and self-esteem. This can have a protective effect against the development of mental health disorders such as anxiety and depression. In contrast, an insecure attachment relationship can lead to feelings of lack of safety, distrust, and low self-esteem, which can increase the risk of developing mental health problems.

Thirdly, as parents, we are key role models for our children, and how we handle stress, cope with challenges, and manage our emotions has a significant impact on our child's emotional and behavioural development (and how our children eventually handle stress and cope with their own challenges in life). This is the mirror neuron system that speeds up our ability to learn, and this is why social connections are so important. For instance, let's reflect a bit here: have you noticed your children behave differently when you're under stress? If you're unsure, next time, observe your children's reactions to your stress; do they mirror your own? When we get good at handling our own stress, our kids naturally learn from us how to deal with tough situations. It's always easier to keep problems at bay in the first place than to fix them after they happen.

When it comes down to it, brain health refers to the well-being of our brains, including our cognitive, emotional, and social functioning. It's about maintaining a healthy brain throughout our lives—from childhood through to old age. So why is brain health so important? For one, our brains are crucial to our daily lives. They enable us to learn, think, feel, and interact with the world around us. When our brains are healthy, we are better equipped to handle the challenges that inevitably come our way. Brain health is not just an individual concern but also a collective responsibility. The well-being of each person in a community or society impacts the well-being of the whole. When individuals struggle with mental health challenges, it affects not only their own lives but also the lives of their families, friends, and communities.

I started my journey into neuroscience because of my sister Francesca's mental illness and the reaction to it from the healthcare system, the broader community, and our family, whose response was layered with secrecy and shame. After 30 years in this field, I learned that educating yourself about brain health is one of the most important tools that offers our children the best start in life and can help prevent and improve treatment of mental health issues. So where do we start to promote brain health as parents and carers? The answer lies in getting to know the brain and its neuroplasticity capability. At the epicentre of brain health is our early life experiences combined with our genetics, which has a *dramatic* effect on brain development and mental health. Our brains are the control centres for our bodies and minds, regulating everything from our emotions to our physical movements. They are incredibly complex and powerful organs. But just like any other part of our bodies, our brains need care and attention to function at their best. And that's where brain health comes in.

In many ways, the brain is like a muscle. Like muscles, it can be trained. While neuroplasticity (the brain's ability to change and adapt in response to experiences and the environment) is possible at any stage of life, the early years offer the best opportunity to break intergenerational cycles, build greater resilience, and help brain health thrive on emotional, cognitive, and social levels. Regardless of our own trauma, we can break the intergenerational cycle by applying practical and simple strategies that, over time, become our own habits and our children's greatest asset.

Many modern parenting books focus on external factors, such as routines and discipline, while forgetting that each child is unique, and that parent–child interactions, such as attention, touch, talk, and play, influence the development of our children's brains—and have a *lasting* impact. This book provides a more individualised perspective on parenting and caretaking and how to bridge the gap between the best advice with the reality and complexity of being a busy and stressed person in today's digital world. It is about how to master parenting using a neuroscientific understanding while presenting everything in an easy-to-digest way so that anyone who wants to do things differently can.

The aim is to support and help you to nurture yourself, your child and others using simple tools that promote brain health using neuroplasticity. In this book, you'll discover:

- "Being Seen" is the key to nurturing secure relationships between parents and children.
- Re-parenting yourself offers invaluable benefits that not only impact your children but also echo through future generations.
- You can master parenting with actionable strategies for addressing challenging topics like technology and intimacy.
- A vital parental and life journey begins by reshaping the overwhelmed and stressed brain to prioritise health and well-being.

Looking back, I wish I had known about the importance of carving out time and attention towards building strong serve and return relationships and learning parenting skills with my children from day one. With this book, I hope to give other parents and carers the tools and resources I didn't have, so they can master parenting in the digital age to then nurture their children's long-term success and happiness. Even though I didn't have all the answers at first, I am proud of the parent I have become and the ways in which I have continued to learn and grow along with my children. In many ways, this book is part of that growth process. Throughout the chapters we'll uncover *simple* and *easy* tools that can help you break free from parenting models that have inevitably been passed down through generations—and tools that won't require huge amounts of energy or effort.

Within these pages, I share findings from research, discussions with countless experts, actionable tips, and more, so you can nurture yourself and your child's brain health potential for life. You will learn a simple and practical approach that focuses on the brain and draws from neuroscience and evidence-based research to help yourself and your child become happier, healthier, and stronger.

At the end of the day, you aren't stuck. You have the superpower to change the direction of your life and your child's future. You don't need to do tons of work. As a parent myself, I understand how daunting adding "more" to an already chaotic schedule can sound. But this isn't necessarily about adding more or feeling guilty when we try but fail. Rather, it's about taking one small step at a time; building our own yellow brick road doesn't happen overnight. We need to lay down one brick at a time. From there, we can watch our beautiful children transform into confident, strong, secure, and kind humans even in the digital age. So, let's get started.

PART 1

The Pillars of Parenting

Chapter 1

BEING SEEN

Understanding Its Importance in Brain and Child Development

In Brief

Parenting is the hardest job on the planet and has become even harder in the digital age. In a complex world, parents need simple solutions for guiding their children into adulthood. Children "being seen" by their parents *is the silver bullet* for building thriving children. The problem we all face in our society is being so stressed and busy trying to put our lives together that we are not able to see our children in the way that builds brain health.

The concept of "being seen" in the parent–child relationship is an intricate experience that goes far beyond mere physical visibility. It involves a parent being fully present and attentive to their child, taking the time to genuinely understand and empathise with them. This commitment includes not only acknowledging and validating their emotions and needs but also taking an active interest in discovering their unique qualities, interests, and talents. By spending quality, uninterrupted time with the child, the parent can better recognise what makes their child individual and special. This deep understanding allows the parent to encourage a sense of autonomy and individuality in the child, while still providing the necessary boundaries and guidance.

Being seen, however, is not a one-time event; it's a consistent and ongoing practice that evolves as the child grows. It's about maintaining a steady presence in the child's life, marked by consistency in emotional recognition, affirming praise, and quality time spent together. This regularity helps to build trust and assures the child that they are not just an afterthought but a valued and integral part of the family unit and society. This long-term commitment is vital for fostering a strong parent–child bond and has a significant positive impact on a child's emotional and psychological well-being.

The concept of "being seen" in the parent–child relationship emphasises the importance of recognising and valuing the child for who they are in the present moment, rather than what the parent might wish or envision them to become. It's not about fulfilling the parent's unmet dreams, carrying on family traditions that may not resonate with the child, or filling gaps in the parent's own history. Being seen involves an authentic engagement with the child's true self, including their unique interests, talents, and aspirations.

Creating an environment of safety and security is not just about physical well-being but also about emotional and psychological safety—it means creating an environment where the child feels free to express their individuality without the fear of judgement or expectation. Consistent emotional support and the freedom to explore their own paths contribute to a child's sense of security and self-worth. In this nurturing environment, the child feels valued not for meeting external expectations but for their inherent qualities. This approach fosters a strong, resilient parent–child bond, allowing the child to feel deeply loved, valued, and securely seen for who they truly are.

Indeed, the societal pressures and expectations on both parents and children to conform to prescribed notions of success can be a significant hurdle in the quest for authentic, healthy parent–child relationships. These societal norms often prioritise external achievements and looks over emotional well-being; through social media we focus on image, grades, athletic prowess, or other easily quantifiable metrics at the expense of holistic development. In such an environment, the notion of "being seen" can be eclipsed by the drive to "be seen *as*" successful, smart, or accomplished in the eyes of society. This focus on external appearances and societal approval can create a disconnect between parents and children, as both may feel pressured to uphold an image that doesn't necessarily align with their true selves.

The challenge, then, is for parents to navigate these societal pressures while still maintaining a focus on the emotional and psychological well-being of their children—and themselves. This involves a delicate balancing act of equipping children with the skills and opportunities they need to succeed in the world, while also protecting space for them to be themselves, free from the weight of external expectations. Achieving this balance is no small feat, especially in a society that often measures worth in superficial ways, but it's essential for the healthy emotional development of both parent and child.

The consequences of failing to balance this understanding can be profound and far-reaching, affecting children's physical and mental health. When the focus is solely on external achievements, driven by societal expectations or parental ambitions, children may experience heightened stress, anxiety, and even depression. The pressure to conform to a prescribed notion of success can contribute to a host of mental health challenges, ranging from low self-esteem and body-image issues to more severe conditions like eating disorders or self-harm and many other mental health disorders.

Moreover, the constant stress and pressure can also manifest in physical health problems, such as sleep disorders, chronic fatigue, and weakened immune systems. In extreme cases, the relentless pursuit of external markers of success can even lead to substance abuse as a coping mechanism. These health outcomes are not just transient issues but can set the stage for long-term health complications that extend into adulthood and affect their lives.

Francesca, my sister, was chosen to play the shark in her end-of-year high school play. She put on an excellent performance, dancing across the stage with a fin attached to her back, wearing grey tights and her signature sneaky, wide grin. The performance had everyone in stitches. When Francesca was on stage, she felt seen for the first time. And this sparked her desire to pursue the dramatic arts, which tragically ended with audiences in mental hospitals rather than in movie theatres.

Francesca's mental health journey and society's reaction to it were mixed with secrets and shame. She was given a typical mental health treatment plan to reduce symptoms with medications, which never resolved the problems she faced. Losing my beautiful sister was one of the hardest moments in my life. They say grief doesn't get easier; we just get used to it. But my hope is to help others avoid this pain by starting at the very beginning and by helping parents guide their children (and themselves) towards improved brain health—and, thus, better mental health that will last a lifetime. My sister's journey into mental illness diagnoses happened before she was born and became exacerbated by her early life experiences of not being seen by parents, grandparents, siblings, and educators, and then being seen in the wrong way by peers, romantic partners, and mental health professionals.

There are enormous costs to children's lives of blindly believing that parenting children is natural and that they are resilient and can overcome adverse childhood experiences (ACEs). "We need to stop believing the lie that young children won't be impacted by what happens in early childhood," says Beth Tyson, a trauma expert and guest on my podcast. These are outdated notions of parenting, and the neuroscience evidence is unequivocal. ACEs contribute significantly to mental and physical health disorders that are diagnosed later in life. There is a chapter dedicated to this important knowledge that needs to be in all parenting resources, pamphlets and pregnancy classes.

As a parent first (who happens to be a neuroscientist), I understand all too well the moments of self-doubt and feelings of failure that parenting can bring. Reading up on new scientific findings often brings back memories of my own challenges as a young parent. There were days when my biggest accomplishment was not smashing the cereal boxes down on the table at breakfast while watching the clock. If I managed to get to work without running a red light or having to pry my child from my leg at day care, I considered that a victory. I'm acutely aware of the relentless pace of life and the difficulties that come with it. As a parent myself, I understand how busy life gets. Between making dinner, juggling a career, ensuring our children are safe, and all other aspects of life that tug at our attention, it can prove difficult to find quality time for ourselves and our children in the demanding times of their lives.

Stepping into parenthood fundamentally changes who you are. For many, it's a pivotal experience that greatly influences our personal evolution. Being a parent is arguably the most challenging job there is, and we all strive to do our best with the resources we have. Our first ideas about parenting typically come from our own parents, passed down like a baton from one generation to the next. Becoming more aware of this legacy involves observing our own reactions when under stress—how do *we* respond to our children in these moments? This awareness is a crucial step in our parenting journey.

In a world increasingly dominated by screens and stressors, the quest for quality time and meaningful interaction becomes an uphill battle. Yet it's precisely within this challenging landscape that the need to be truly "seen"—for us and our children—becomes paramount. Stress and smartphones may be an inevitable part of modern life, but their impact on our brain and that of our children is not set in stone. Through deliberate efforts, we can learn to manage stress and screens effectively, not just for our well-being but as a model for our children, whose brains are at a highly plastic and formative stage.

"Being Seen" by Parents and Caregivers

Have you noticed and wondered why children always look for attention from their parents or carers or educators—or anyone? You can easily see this in any playground; "Look at me, Mum!" "Mum! Dad! Watch this!" Most scientists agree that this is an evolutionary drive that keeps us safe from birth to around age three. But it also teaches us the basics of life, allowing us to mimic what our parents do. We are learning what it means to become human, physically and emotionally. And we are learning the basics of human interaction and what a healthy relationship means.

Learning the parenting skills of "being seen" is the least understood and most underestimated yet critical aspect of child development, particularly within the family context. Being seen is more than providing a physical presence; it is paying attention to and responding to a baby and child and adolescent. During the first few years of life, an infant's brain is wired to grab attention. This is not merely a behavioural attribute but a critical part of its ability to survive. Because babies may not have words to communicate, they are incredibly skilled at using a variety of tricks to capture adult attention. These tricks serve as fundamental survival mechanisms, ensuring that their basic needs such as food, water, sleep, and safety are met. Not only are these behaviours instinctual but they are also supported by the developing neural networks in a baby's brain, making them highly effective.

The most obvious and well-known method babies use to grab attention is crying. Crying serves as a universal distress signal that is hardwired to provoke a response from adults. Different cries can even indicate different needs, from hunger to discomfort, allowing caregivers to respond appropriately. Neurologically, the sound of a baby's cry is designed to be attention-grabbing and emotionally affecting, triggering caregiving instincts in adults. The baby's brain, at this stage, is in a period of rapid growth, with millions of neurons forming connections at an extraordinary pace. These connections, known as synapses, are influenced by external stimuli, and attention from caregivers is among the most potent forms of such stimuli.

Grabbing attention extends beyond mere survival; children's brains are like sponges, soaking up the actions, words, and even the emotional states of people around them. Think of their brain as a supercomputer that is updating its software every moment it learns something new. What they witness becomes a part of their brain development and educational fabric, stitched together over time to form their understanding of the world and their place in it. Adults are the most powerful instrument in guiding a child's brain development, mind, behaviour and life.

When a child's emotional states are being seen by their parents and caregivers—through eye contact, touch, or verbal acknowledgement—it triggers a cascade of neurochemical events. The release of neurotransmitters like dopamine and oxytocin not only makes the child feel seen but also reinforces neural pathways associated with positive emotional experiences and feelings of safety and security. This phenomenon is part of the brain's plasticity—its ability to change and adapt, which is most pronounced in the early years. When a child's feelings or actions are validated, certain neural pathways that contribute to a healthy sense of self are reinforced. This neural reinforcement helps the child develop emotional resilience and self-esteem, which are key elements for mental health. For this reason, being seen in the right way through childhood by parents and carers is a key ingredient for building thriving minds and children. Understanding and practising the parenting skills that ensure children feel seen are among the most lasting gifts we can offer. However, this is difficult to deliver in the real world as we juggle the complexities of modern family life.

Children learn by copying their parents. I saw a demonstration of this recently; I was sitting on a bus, quietly looking out the window, when a young mother and her toddler took the seats next to me. After a moment, she reached into her handbag and pulled out a lipstick. Without taking her eyes off the screen, she skilfully applied the crimson hue to her lips. I couldn't help but marvel at her ability to balance technology, the bus turning a corner, and motherhood seamlessly. Then, a tiny giggle caught my attention. The toddler, sitting next to her, with her small hands, held a doll up to her face and pretended to apply lipstick, miming the exact movements her mother had just performed. She even made an approving nod to her doll.

This seemingly trivial moment was, in fact, profound. It illustrated an age-old educational concept that actions often speak louder than words. Whether it's a toddler emulating her mother or a student absorbing the behaviour of a teacher, the act of imitation is often our first step into the world of learning. This is how a child learns in the first few years of life: by copying their parents, carers and educators. In this digital age, the implications of this are highly significant.

The long-lasting impact of feeling seen is profound. Research indicates that children who experience consistent emotional recognition and validation have healthier brain development, better stress regulation mechanisms, and higher emotional intelligence. These neurobiological benefits are lifelong and form the basis for emotional resilience and mental well-being in adulthood. Every action we perform, whether consciously or unconsciously, serves as a potential lesson for someone else. It emphasises the responsibility we have, not just to impart knowledge but also to demonstrate values, ethics, and skills through our actions.

We grew up in Nanango, a small country town in Queensland, where Mum and Dad started their small community pharmacy that they would run for the next 40 years. Francesca was one of four children under the age of seven. There was no extended family, and women were expected to be domestic goddesses while the fathers were hard at work. Our parents were barely adults forging a new life in a strange place. We were oblivious to any of this, outside racing our motorbikes up and down the country lanes, giggling, and being naughty, building cubby houses in the hoop pines, and having fun with the plethora of children on the street.

Chapter 1: BEING SEEN

Francesca was the second born, sandwiched between me, the eldest, and my brother and before my youngest sister. Mum was pregnant with my brother within two months of having had Francesca (and this possibly took away some of the attention Francesca needed in those early years). We roamed and explored the neighbourhoods and were in and out of other people's homes at whatever time we wanted; we just had to come home for dinner. This felt normal until we went to high school in the Big Smoke; some of our peers seemed to be more well-read and articulate than us. On the outside, I looked just like any other student, but inside, I felt different. Catching the city bus, reading tough books, and writing long, tricky sentences were all new to me. I used to feel free, but now, it was like I was trapped in a place I didn't quite fit into.

In Grade 3, Francesca was locked up in a dark room for hours by a nun in her school as a punishment. At the time, we were children, and we didn't understand how damaging this could be. It was not until I had my own children and was in my forties that I realised that the stressful experiences in her early life had played a major role in Francesca's mental health journey. She was the funny one, getting in trouble and being expelled from school; then in boarding school, managing to have all the year 12 students sneak out at midnight and go to a nightclub. After school, she started meeting the wrong people. We were living together; I was 20 at the time and had no idea she was hanging out with a much older crowd in the Valley in Brisbane, except when one day, she arrived home at 4 am on the back of a motorbike. I had to leave at 7 am to do an anatomy exam, so nothing registered except that Francesca was being naughty again. I hid it from Mum and Dad. I did not want it to be seen.

One afternoon, she disappeared from Brisbane. At 18 she took her clothes in garbage bags and moved to Sydney to try to become an actress. She ended up working for a famous talent agent in Sydney at the reception desk. It was here she met the love of her life. One time, she came home wearing his leather jacket, and she was so happy. The next time she flew home alone and in great distress. We were lucky that a friend in Sydney had managed to get her on the aeroplane. Many people have heartbreak but do not end up in a mental health crisis as severe as Francesca's. The sum of her early life experiences coupled with traumatic events in Sydney changed Francesca's life and sent her brain into a meltdown. When she received her initial diagnosis in the hospital in 1989, the question "I wonder what happened to her?" wasn't raised by anyone. We were all in the dark about the events leading up to this moment.

Imagine if we had seen Francesca through her life experiences and had seen her story rather than jump to a rapid diagnosis that led her to a lock-up ward and medications. If it had been illuminated on a billboard? She had many adverse experiences that were compounded by genetic history. I share her story to encourage you to see the people in your life for the first time and not focus on their symptoms only. You will begin to see their strength and resilience. Francesca even said to my other sister and me that it was lucky that *she* ended up with schizophrenia because *we* would not have handled it.

It is not possible to unsee the neuroscience of early life experiences and trauma and their role in mental health disorders later in life. The research provides undeniable facts. After spending decades trying to understand the brain, for me it was a hard moment to see these effects expressed in my family history. But, like all knowledge, the research raises awareness and education. Early life experiences, including trauma or adversity, play a crucial role in impacting brain function. The effects of these experiences can differ among individuals, which is why some may develop mental health disorders while others might not. In essence, our early experiences can influence the way our brain processes emotions and handles stress, which can impact our mental well-being later in life.

By 2005, Francesca still had not been seen except by a lot of healthcare professionals; she was isolated from the community and only occasionally spent time with her family. She was on multiple medications, as is the tragedy of many individuals with mental health disorders today. Surprisingly, new research is emerging, indicating that maybe the value of pharmacotherapies has been overestimated and their shortcomings have been underestimated. This is a newly studied topic, one which I hope receives more attention in the coming years. Medicines are valuable in some instances and can help in acute and short-term situations. However, very few of the medicines have been examined for use over long periods of time as they are now being taken.

Most of the clinical studies examined people using the drugs for up to 12 months. I am a pharmacist and understand well the benefits and side effects of medicines. The benefits are always touted over the side effects. It is unacceptable for the public to think that medicines alone can reduce the impact of early life experiences on brain function that leads to mental and physical health disorders.

At this point, Francesca had now *become* her illness and medications and, ultimately, there seemed to be no way back. Here's a short list of what happened to Francesca on the drugs. I discuss this to raise awareness of consequences of people taking these types of drugs for long periods of time. She had a permanent tremor in her hands; her pinky finger ended up poking to the side rather than straight ahead. She became obese and developed metabolic dysfunction that could not be reversed. Her symptoms gradually worsened, so they escalated the dosage of the drugs until she was no longer functioning in the world. On one of her drugs, she had to have regular blood tests to make sure she did not develop serious blood conditions. On the highest dose, she had a grand mal seizure; she survived because a stranger found her on the ground in the middle of a park and pulled her tongue down to allow her to breath. At the end of her life, she was on the highest doses of a rotating group of antipsychotic drugs.

This is really important information for us to absorb and understand, especially as younger and younger children are being given drugs. I know, as a pharmacist and a neuroscientist, that these medicines are potent and provide short-term fixes in some cases. Drugs are potent chemicals that change the brain and possibly the individual—this is what I observed with my sister over 15 years. Being seen by our parents or carers or at least one person who has our back in a supportive and caring society is as important as or more important than breastfeeding and early life education. Having the time and ability to be fully present for your child, both physically and emotionally, promotes their healthy brain development and emotional well-being, and should never be underestimated.

As a society we must understand that children need quality time and attention in the early years of life from healthy adults supported by healthy communities. It is possible to outsource this essential ingredient, but that has long-term deleterious effects on children. Caring for children is time consuming in the early years, but as someone who has experienced the impact of mental health disorders, I can assure you that not caring well for children impacts the family for the rest of their life and prevents them from reaching their potential.

It's not fair to blame just one person or one family for mental health problems, because the conditions that contribute to them are societal and have been entrenched in our communities for a long time. The problems are too much for just Mum and or Dad or carers to fix on their own. We all need to work together—everyone, no matter who they are—to make things better for our kids and all the kids who will come after them. This situation underscores that parenting goes beyond just material provisions; it also involves giving time, attention, and a consistent presence to build strong, loving relationships with our children.

This is the absolute bedrock of active listening, empathetic understanding, and a commitment to foster an environment where both parent and child feel safe, calm, valued, and heard. As we navigate the challenges of screen time, work pressures, and other stressors, let's not lose sight of the fundamental need we all have for genuine connection.

It's no secret that mental health issues are on the rise, and social media is undeniably a major player in this uptick. Studies have shown a concerning correlation between excessive smartphone use and various mental health issues in children and adolescents. The constant comparisons with others and seeking of validation can have adverse effects on a child's self-esteem and mental well-being. From anxiety, depression, eating disorders, addiction and self-harm to sleep disturbances and attention problems, the impact of smartphones on young minds cannot be overlooked.

In 2023, about 60% of the world's population is on social media, and these apps steal our attention for an average of 2 hours and 24 minutes a day. Our attention is now being diverted to screens, away from children, and children's attention is being channelled to tech and devices. When using these apps, the reward centre of our brain lights up the same way it does when we use nicotine or alcohol. In other words, social media is just as addictive as some of the most dangerous drugs on the planet. We are now being seen by screens rather than people.

And yes, this should alarm any parent or carer. There are no two ways to talk about this. Social media addiction and its negative impact on brain health and mental health is real. And it's happening right now. It has a detrimental impact on the developing brain—something all of us, as parents, should be aware of, but most of us aren't.

Therefore, the importance of equipping parents with the knowledge and educational resources to foster an environment where children feel securely seen for who they are cannot be overstated. The aim should be to cultivate a strong parent–child relationship that prioritises safety, security and mental health, as these are the foundations upon which a thriving and fulfilled life can be built.

Making Time and Every Moment Count with Our Kids

While writing this book, I discovered one key problem: most parents don't have *enough time* to sit down and read. The feedback I got from parents was that time seemed to be the biggest hurdle. I shouldn't have been surprised, as I remember all too well barely having a minute or two for myself in those early days as a new parent. Life changes a lot when you go from having no kids to becoming a parent. Here's a simple breakdown of how people usually spend their day before and after kids come into the picture:

Before Having Kids:

1. **Basic Stuff:** Think of things like sleeping, eating, and taking care of yourself. Most people might sleep for about 7–9 hours, take 1–2 hours for meals, and another hour just for themselves. Total = 9–12 hours
2. **Work and School:** Based on their job or studies, people might be busy with this for about 7–9 hours a day.
3. **Fun and Friends:** Hanging out with pals, enjoying hobbies, or just relaxing. Without kids, people have a good 3–6 hours daily for this.
4. **House Stuff:** Things like cleaning up or shopping for groceries might take up 1–2 hours daily.

After Having Kids:

1. **Taking Care of Little Ones:** With babies, especially, parents might be busy for a whopping 8–12 hours (or even more) doing things like feeding, changing nappies/diapers, or just calming them down.
2. **Less Sleep:** With a baby around, uninterrupted sleep becomes a luxury. Parents might squeeze in 5–7 hours, with a few wake-up calls in between.

3. **Work Time:** Maternity or paternity leaves, or choosing to stay home, can change how much time is spent working. Plus, there's the juggling act of fitting work around the baby's routine.
4. **More House Stuff:** There's more laundry, cleaning, and cooking with a baby in the house, maybe taking up 2–4 hours daily.
5. **Less "Me" Time:** Free time and hanging out with friends might reduce to around 1–3 hours because, well, baby comes first.
6. **Taking Care of Yourself:** Even with all the baby chaos, it's important to find a little time, even if it's less than before, for things like a quick workout or just some relaxation.

But in order to make this book as useful to parents as possible, I realised I needed to help parents solve the problem of time. Take a "T.I.M.E." audit to help you leverage your most precious resource and pinpoint where improvements can easily be made. So, let's dig in.

T – Track. Begin by tracking your time for a week. Note down what you're doing in a day. This gives you a clear picture of where your time is going and will help you identify patterns.

I – Identify. From your tracking exercise, identify which activities align with your family's goals and values, and which ones don't. This step helps you to recognise what is truly important and what might be taking up valuable time unnecessarily (such as social media).

M – Modify. Based on what you've identified, modify your schedule. Let go of activities that aren't serving your family's purpose and make room for those that do. This could involve less screen time and more quality family interaction, less multitasking, and more focused, meaningful engagement.

E – Educate. As you're going through this process, involve your children. Use this as an opportunity to educate them about time management and priorities. Show them how you decide what's important and how you organise your time accordingly. By doing so, you're equipping them with vital life skills they will use for the rest of their lives.

Remember to reassess regularly, as needs and priorities can change. The "T.I.M.E." audit technique isn't a one-time exercise but rather a living process that evolves with your family's journey. It's a simple yet effective way of managing time and ensuring it aligns with your family's overall direction.

Remember, our goal isn't to cram more into each hour, but to ensure that our hours are filled with more of what truly matters. By developing a keen consciousness of time, we can move from merely spending time to investing it in meaningful ways and creating a family life and environment that fosters safety, calm, and deep connection. In the fast-paced world of instant gratification, the allure of quick returns can often overshadow the merits of long-term investment. But just as the tallest trees take years to reach their peak, the most enduring successes are often built on sustained effort, patience, and a vision that extends beyond the immediate horizon.

Time well spent is not about mere quantity but quality. It's about recognising the value in planting seeds *now* that will yield results in the future. Investing time wisely involves a deeper understanding of one's goals and the steps required to achieve them. This may mean foregoing immediate pleasures for bigger rewards down the line. Consider, for example, the choice to invest in one's education or personal development. While the immediate benefits may not be readily apparent, the long-term gains in terms of knowledge, skills, and career opportunities can be immeasurable. Similarly, the decision to spend quality time with loved ones, nurturing relationships, and building bonds might not yield immediate tangible results but can lead to deeper connections and memories that last a lifetime.

When my children were young, finding time was like locating a needle in a haystack. It's not easy. And the more parents I talk to, the more I realise that most parents are short on our most valuable resource: time. Yet our time and undivided attention are such important ingredients for creating thriving children, especially in the modern world—and the research shows that on average, we have about two spare hours every day. In a rapidly evolving digital landscape, where screen time often competes for family time, the importance of quality time and attention in parenting cannot be overstated.

When we become acutely aware of the passage of time, and, more importantly, how we use it, we can make space for the things that truly matter. Now, think of this compass as a tool to help declutter your days, to shave off the non-essentials and anchor what truly matters. Parenthood invariably reshapes our lives, and things we once took for granted—like leisurely Netflix binges or endless social media scrolling—might need to take a backseat. And that's okay. The joy of raising children requires certain sacrifices, but remember, these are temporary.

Try the T.I.M.E. technique to solve the time puzzle. Consider your routine and the elements that enhance the relationships with your child. Focus on those activities that cultivate connection and let go of those that don't. And don't forget to prioritise sleep! It's not just a rejuvenating break; it's a critical ingredient in your family's well-being recipe. Swap that late-night screen time for sleep, and just imagine the transformation that could spark in your home all by itself. Feel the energy levels rise, patience expand, and stress dissipate. Now, that's a change worth making! So, let's harness the power of time, aligning our actions with our deepest values. Let's turn the notion of being time poor on its head and instead cultivate a richness of meaningful, connected moments.

Perhaps this means reshuffling daily routines to prioritise uninterrupted family time. Maybe it's about adopting practices, morning routines, exercise, and sleep (which we discuss in the following chapters) that allow us to be more fully present during our interactions with our children. It could even involve creating new family rituals that foster connection and mutual respect. The specifics of your compass will depend on your family's unique needs and circumstances.

Parenting is like weaving a beautiful quilt, where sometimes the quiet moments, like truly noticing our kids, create the warmest patches. Every kid wants to be seen not just for the big stuff but for who they really are, with all their happiness, worries, and dreams. By learning more and being there for them, we can make our home a place where our kids feel truly loved and appreciated. In a world that's always pushing and pulling, we need to be the constant voice telling our kids that we are there for them no matter what. Feeling loved and accepted helps them become confident and resilient. But this isn't always easy for us as parents, especially if we didn't get that same love and support from our own parents.

Often, without realising it, we might find ourselves doing the same things that our parents did. It's like a pattern passed down through generations, a dance that keeps repeating. To get started on making the change, take just five minutes from your day to spend with your child. Even a short moment of your undivided attention and affection can mean the world to them. ***Remember, it's not about doing it all; it's about making what you do count.***

Time for Reflection

We often find ourselves caught in the whirlwind of being busy with daily tasks, obligations, and digital distractions. As the days blend into weeks, we may wonder, where does our time truly go? Start by charting out your day on a piece of paper, with distinct columns for each activity, the time you dedicate to it, and your satisfaction level ranging from 1 (least satisfying) to 5 (most satisfying). This chart is not just a list; it's a mirror reflecting on how you spend your time during the day—for example, in morning routines, time with the family and friends, work, screen time, and even those sneaky social media scrolls. Then, ask yourself: What drives me to these activities? Is it duty, pleasure, habit, or something else? Do they uplift me or weigh me down? As clarity emerges, you'll find the power within to realign, making choices that not only meet obligations but make time for what truly matters.

Chapter 2

SEEING YOURSELF AS A PARENT

Cultivating Self-Awareness

In Brief

Picture this: you've got a magic lamp, and every time you rub it, out pops a genie ready to help with today's parenting challenges. What would be your first question? Maybe you'd ask how to get the right balance between your kid's screen time and playing outside. Or perhaps you'd want a heads-up on which new gadgets are good for them, and which ones you should pass on. Such a genie wouldn't just grant wishes; it could serve as a soothing voice during those late-night moments when you're questioning your decisions, or as a confidant when you're feeling overwhelmed by the unceasing complexities that technology brings into the home.

The one factor that most parents can universally agree on is the desire for their children to be healthy and happy. This fundamental wish transcends cultural, economic, and social differences and is often the primary motivator behind a wide range of parenting decisions. Parents and caregivers play a special role in helping their kids grow up healthy and happy. They help shape the way a child's brain develops. But raising kids is a big job, and it's not something parents should have to do all by themselves. They need help and support to give their kids the best start in life.

Let's take a moment to acknowledge a critical truth: to help others thrive, including our children, we first need to put in place actions to help *ourselves* thrive. This means that we must go beyond our roles as parents, spouses, caregivers, or whatever other titles we hold. It's akin to the aeroplane safety rule of putting on your oxygen mask first before helping others.

Contrary to popular belief, we aren't born as blank slates. Our initial mental frameworks—our beliefs, values, and even our styles of parenting—are often inherited from our parents and past generations. Rather than going in circles with old habits, think of summoning your genie to shed light on how you think and the biases you might have, especially in your role as a parent. With this newfound understanding, rewrite your parenting journey. Dive into understanding your brain, and let that knowledge guide you to face parenting with a newfound confidence.

Parenting Experience Changes Your Brain and Life

Social media influencers often paint a rosy picture of parenthood, suggesting that your pre-baby lifestyle can easily be maintained after welcoming a new child into your life. While these curated images and stories may be aspirational, they often don't reflect the complexities and adjustments that come with having a baby. Parenting is a transformative experience that usually requires significant changes in daily routines, priorities, and even personal goals. The notion that one can simply "add a baby" to an existing lifestyle without making any substantial changes is not only misleading but can also set unrealistic expectations for new parents, leading to unnecessary stress and disappointment.

The day my babies were born, on each occasion the hospital room was filled with a mixture of emotions—excitement, anticipation, fear,

Chapter 2: SEEING YOURSELF AS A PARENT

and love, all swirled into a momentous bubble that seemed to envelop everyone present. As the first cry echoed through the room, it was as if time paused, allowing us to soak in the gravity of the moment.

The initial weeks were a blur of sleepless nights, nappy changes, and countless Google searches like "Is this normal for a newborn?" or "How to soothe a crying baby". Our previous identities seemed to dissolve into this new role that neither of us fully understood yet. Our conversations revolved around feeding schedules, and our social life took a backseat. At around the three-month mark, amidst a particularly challenging night of ceaseless crying, it struck me—am I the same person I was before the baby was born? The answer was both simple and complex. No, I wasn't. The person before would have been overwhelmed by the very thought of such responsibility. But the person I had become found strength in that responsibility. I had a newfound respect for my own parents and a deep sense of empathy for all caregivers around the world. I looked at the tiny, fragile life in my arms and realised that from this day on, everything would be different.

Are you the same person as you were before a baby? Definitely not. The philosopher L.A. Paul describes becoming a parent as a quintessential example of a life-altering transformation. Before you experience parenthood, it's virtually impossible to fully grasp its multifaceted nature. The arrival of a child often prompts a fundamental shift in your core values and life objectives, reorienting them around this new, paramount concern: your child. Your emotional landscape is also profoundly altered, with new forms of fear and happiness entering your daily experience. This happens because of the biological need the child has to gain your attention, for the sake of survival.

The transformative journey of parenthood does more than introduce new forms of joy and challenge; it often makes memories and emotions tied to one's own upbringing resurface. The act of parenting can trigger reminders of past experiences, both pleasant and traumatic, creating an emotional tapestry that is as complex as it is vivid.

These flashbacks can serve as both cautionary tales and inspirational blueprints, depending on the nature of your own childhood and the relationship with your parents. Thus, parenthood doesn't just redefine you in isolation; it also reshapes your understanding of familial relationships and generational narratives. It's a profound transformation that affects not just how you see your child and yourself, but also how you view your own parents and your place in your family history. As neuroscience continues to grow and expand, there's been an emergence of a new field called "parental neuroscience". While this field is still evolving and in the very early stages, recent findings suggest that becoming a parent significantly alters brain functioning and how we think, feel, and behave. I was not just a person anymore; I was a parent.

Knowing and understanding ourselves as people and as parents can help us make changes in the right direction to enhance our own brain health and that of our children. Parenting is not a one-way street; it profoundly affects both the parent and the child at a neural level. As we reflect on the stages of brain development, we also acknowledge that no parent is perfect. We learn from our mistakes and grow alongside our children. It's essential to forgive ourselves for our imperfections and embrace the journey with compassion and understanding.

Becoming a parent brings about profound changes, not just in lifestyle but also in brain structure and function. For instance, parenthood is associated with structural changes in the brain, including increases in grey matter, which is vital for social processing and empathy. Hormonal changes, particularly rises in oxytocin and prolactin levels in both mothers and fathers, enhance parental bonding and caregiving behaviours. New parents often face sleep deprivation, which affects brain function and subsequently impacts mood, cognitive abilities, and decision-making. Emotional regulation is another area where the brain adapts, as specific regions are more activated in parents, possibly to better respond to their child's needs and to form

Chapter 2: SEEING YOURSELF AS A PARENT

deeper emotional bonds. Parenting also stimulates increased brain activity in regions associated with social cognition, which might be due to the complex social cues involved in caregiving.

Furthermore, becoming a parent can cast your own parents in a new light, compelling you to revisit and reevaluate your perceptions of them. Their choices and actions, which may have once been sources of confusion or resentment, may take on new meaning as you navigate similar challenges and triumphs. Suddenly, their struggles and sacrifices might appear more palpable, their successes more admirable, or their failures more understandable. The role you now inhabit brings forth an intricate blend of empathy, critique, and introspection regarding your own parents, enriching your understanding of the intricate dynamics that shape family life.

Understanding these changes can help us navigate parenthood and allow us to achieve a higher state of brain health while guiding our children towards success. Finally, it's important to note that the brain retains its plasticity throughout life, meaning it can change based on experiences; thus, parenting undeniably impacts both brain structure and function.

Think of your brain as your personal supercomputer. Learning to master this incredibly adaptable organ can be one of the most valuable skills you acquire. It's not just about becoming more efficient or intelligent; it's about enhancing your overall well-being. This mastery also equips us to better handle the complexities of modern life, from navigating technology and stress to fostering healthier relationships and having more confidence to parent children growing up in the tech age. The journey to change, marked by increasing neuroplasticity—the brain's ability to reorganise itself—begins by understanding our own minds and the role of our own parents and grandparents.

Where did your brain and parenting style come from?

How does the brain and parenting style evolve over time?

How does your approach to parenting shape the brains of your children?

Once we understand that we are not confined by our initial mental and parental frameworks, the possibilities are endless. Whether it's acquiring a new skill, expanding our understanding, or overcoming challenges, the potential for transformation is immense.

Where Do "We" and Our Parenting Style Come From?

Many of us, when becoming parents, aspire to do things differently from how our own parents did them. We ponder this during the pregnancy or even before, vowing to avoid certain mistakes or to embrace methods we wish had been applied in our upbringing. However, the reality of parenting—especially when under stress—often tells a different story. Stress has a way of triggering deeply ingrained neural pathways, directing our actions and reactions in ways that are almost automatic.

 CASE STUDY – Tina

Tina sat in the cosy corner of her nursery, a room filled with soft pastels and whimsical decor, staring at the parenting books that lined the shelves. She was eight months pregnant and had devoured them all, each page offering a promise of enlightened parenthood. *I will be different*, she thought, recalling her own childhood marked by inconsistent discipline and a lack of emotional support. *I'll be the parent I wish I had.*

Weeks later, baby Ruby arrived, and Tina felt an indescribable joy mixed with an overwhelming sense of responsibility. The first few days were a blur of feeding, nappy changes, and brief moments of sleep. Despite the exhaustion, Tina was committed to her new-age parenting approach—gentle discipline, open communication, and lots of emotional validation.

However, reality set in as the weeks turned into months. One afternoon, Ruby, now a toddler, threw a tantrum in the middle of the grocery store. Overwhelmed and sleep deprived, Tina felt her stress levels skyrocket. Before she knew it, with quick, jerky movements she grabbed her arm and screamed the dreaded words, "Oh, great, that's just what I needed", and then, "You're fine, stop overreacting!"

After a few seconds, Tina was immediately transported back to her own childhood, remembering the countless times her mother had exploded in a similar way. *What just happened?* she thought, feeling a mix of guilt and confusion. *This isn't me; this isn't how I planned to parent.* Tina realised in that moment how deep the neural pathways of her upbringing had been ingrained. Despite her best intentions and extensive reading, stress had triggered an almost automatic response that mirrored the parenting style she had experienced. It was as if her brain had hit a "default" button, bypassing all the new techniques and philosophies she had so eagerly adopted.

That night, after putting Ruby to bed, Tina returned to the corner of the nursery and picked up one of her parenting books. As she flipped through the pages, she understood that knowing what kind of parent you want to be and actually being that parent were two different things. Real change required not just intention but also awareness and practice. Tina began to understand that parenting was not a linear journey but a complex tapestry of old habits, new aspirations, and countless unpredictable moments. She realised that the road to becoming the parent she aspired to be was

> filled with bumps and detours. And, most importantly, she learned that it was never too late to change course.
>
> From that day on, Tina adopted a new mantra: "Be aware, be patient, and be kind—to Ruby and to myself." And with each stress-filled moment that tested her resolve, she reminded herself that the neural pathways of old could be overwritten, but only if she consciously chose the new path, again and again.

These neural pathways are shaped by years of experience and observation, beginning as early as our own childhoods. When we're stressed, the brain often reverts to these well-worn paths, even when we've intellectually committed to a different approach. It's akin to driving on autopilot; before you know it, you've taken the same old route home even though you'd planned to try a new one.

The neural pathways related to parenting are particularly deep-rooted, likely because they've been reinforced by strong emotional experiences and societal expectations. They not only dictate how we parent but also how we feel about our parenting. This is why, in moments of stress, you may find yourself raising your voice despite having read all the articles about the benefits of calm communication. Or why, despite your best intentions, you find yourself resorting to methods of discipline or emotional support that mirror those of your own parents.

Understanding that these deeply embedded neural pathways influence our parenting, especially under stress, is the first step towards making a conscious change. By recognising these automatic reactions, we can begin to replace them with the actions that align with our parenting ideals. It's not an overnight process, but with awareness and practice, we can rewire those neural pathways to be the kind of parent we aspire to be.

Chapter 2: SEEING YOURSELF AS A PARENT

So, where do we begin? Right here, by taking the time to understand the organ that governs it all: the brain. In the coming sections, we'll delve deeper into neuroscience, offering practical insights and strategies to optimise brain function for both ourselves and our children. So, what does it mean to say that the brain doesn't start as a blank slate and that we revert to our own parents' style? Well, our genes determine some of it, but so does our environment. With some of the foundations already in place, the first few years of our lives drastically impact how our brain develops. Our environment plays an undeniable part in life outcomes and behaviours and can even significantly alter gene expression.

Genetics is the study of genes, which are the units of heredity that are passed from parents to children. Each individual carries two copies of each gene, one from each parent, which interact in complex ways to create a unique individual. These genes are composed of a code called DNA and are located on chromosomes within the cell nucleus. They serve as the blueprint for the construction and functioning of an organism, providing instructions for the growth, development, and maintenance of all biological systems.

Genetics is like the original recipe book for how we're built, written in DNA. When a baby is made, it gets a mix of genes from both Mum and Dad, which helps determine everything from eye colour to personality traits and even how likely they are to be good at sports or music. When it comes to how a child's brain develops or what kind of person they'll grow up to be, genes are definitely part of the story. Just as some families have a history of being tall or good at maths, some families might have a history of being more prone to stress or certain health issues.

But here's the kicker: genes aren't the whole story. Imagine your genes are like the musicians in an orchestra. They can make music on their own, but they sound a lot better when they're conducted well. That's where parenting and the environment come in. The surroundings in which a child grows up, including their home, school, and community, have a significant impact on brain development. Safe, calm, stimulating, and nurturing environments contribute to better outcomes. Childhood maltreatment or ACEs can impact brain development negatively. The experiences you give your child, the food you feed them, the way you support them emotionally—these are like the conductor guiding the orchestra to make beautiful music.

So, even though you can't change the instruments (genes) your child was born with, you can help them play their best music by creating a loving, stimulating, and healthy environment. It's this teamwork between the genes and the way you raise your child that helps them become the best version of themselves.

Did you know that your baby is born with the ability to recognise faces, which is important for social interaction and bonding? Or that they can already tell their mother's voice apart from others'? In fact, studies have shown that just two days after birth, babies already prefer looking at faces over other objects. Watch your baby respond to your facial expressions and voice. They are *learning* the minute they join this world. As you see them grow and develop, take time to appreciate this incredible and exciting journey of discovery! (In my opinion and many parents' opinions, this time ends far too soon, and we often end up reminiscing over these moments of joy and bonding, so take it all in while you can. The human experience is truly spectacular!).

During the early years, our brain undergoes rapid development, influenced by both positive and negative experiences, which together lay the groundwork for our mental health and well-being in the future. Generally, "healthy brain development" refers to the typical growth and changes that occur in the human brain from birth to early adulthood (around the age of 25 years). It's within this timeframe that the brain forms an abundance of new connections, refines existing pathways, and undergoes significant changes in structure and function. With the right parenting, environment, and more, the development of the brain tends to follow a predictable sequence of events, with different areas of the brain developing at different rates. For instance, the prefrontal cortex region of the brain, which is responsible for decision-making and impulse control, continues to develop into early adulthood. Unlike some other regions of the brain, such as the brainstem and midbrain, this area doesn't just finish developing within a condensed timeframe.

When it comes down to it, brain development is undeniably a complex and dynamic process that is impacted by our genes, the environment, and our experiences. When we understand this process, we can learn to provide the necessary support and stimulation to our children during those early years, as well as, perhaps, learn how our own experiences, genetics, and environment shaped our brains—and, thus, our behaviours, triggers, and reactions.

How Do the Brain and Parenting Style Evolve Over Time?

Think of the brain as a supercomputer and an endlessly adaptable, ever-changing landscape—like a city that never sleeps. Picture a brand-new, bustling metropolis, with construction crews feverishly laying down roads and erecting skyscrapers. These are your neural connections in your brain during babyhood, building the complex network that will become your thoughts, your movements, even your favourite foods. This is your brain in hyperdrive, setting the stage for all the learning and growing you'll do in your lifetime.

Now, fast forward to your teenage years. Imagine your cityscape undergoing a major renovation. The City Hall of your brain, known as the prefrontal cortex, is getting an upgrade. This is the control centre for all your decision-making, your "think-before-you-leap" moments. But watch out! The emotional neighbourhoods in your brain are like party districts with neon lights—they're more alive than ever. This explains why teenagers can be a whirlwind of impulsive decisions and emotional roller-coasters.

Think of nutrition as the quality of the building materials. The bricks and mortar, if you will. Providing a steady supply of nutritious food, especially during those early years, is like sending top-grade materials to a construction site. Swap out those ultraprocessed foods and sugary drinks for the high-quality stuff, and watch your city flourish with sturdy, well-designed buildings. Physical activity acts as the city's public parks and recreation areas. It's not just about aesthetics; these green spaces are crucial for the city's well-being. Exercise improves the brain's infrastructure, bolstering cognitive function and emotional health at every age.

Now, let's talk about the city's community and social interactions. Imagine neighbourhoods where people are kind, supportive, and socially engaged. Such a positive atmosphere doesn't just make for happy residents; it also influences the city's ability to handle stress and adversity. A well-connected community enhances emotional regulation and cognitive abilities in its citizens. In today's digital era, technology is like the city's internet infrastructure—both a blessing and a curse. It can facilitate amazing advances, connecting different parts of the city like never before. Yet too much screen time can lead to congested information highways, affecting attention span, emotional balance, and even the city's sleep cycle.

So, here you are, at the helm of this ever-evolving city, watching it grow and change through the first 25 formative years and beyond. It's a thrilling, challenging, and deeply rewarding endeavour, one that requires a mix of planning, adaptability, and a whole lot of love. As you watch your city—your child—develop, remember that every decision you make, whether perfect or flawed, contributes to this incredible, ongoing journey of construction and reinvention.

As you age, your internal city doesn't grow stagnant; it's not a forgotten ruin but a living, breathing entity. Sure, some buildings may show wear and tear, and you might forget where you put your keys, but your city is still capable of change. Your brain remains a hotbed of construction activity, reshaping itself based on new experiences, learning, and memories. Old dogs can learn new tricks, and your brain's ability to adapt—its neuroplasticity—is your lifetime ticket to resilience, wisdom, and new adventures. So, from cradle to cane, your brain is a marvel of construction and reinvention, continually reshaped by your experiences, your emotions, and even your choices.

How Does Our Approach to Parenting Shape the Brains of Our Children?

From the moment we're born until we become adults, our brains are always changing. In the first couple of years, really important changes happen fast. As we keep growing, our brains keep changing too, but in different ways. In the past, scientists could only look at the big picture of how our brains grow. We did not have the technology to be able to see inside the brain, which led to a lot of stories and examination of *behaviour* rather than understanding what was going on *inside the brain*. But now, thanks to special brain scans called MRIs, scientists can see even tiny details of what's going on inside.

The Magic of Modern Brain Scans

These MRIs are like super-powerful cameras that let scientists see all kinds of things about our brains, like how the cells are organised and how different parts of the brain "talk" to each other. What scientists have found is that in the first three years of our lives, there are fast changes in the "white matter" of our brains. That's the part of the brain that helps cells communicate with each other. These changes probably occur because the cells are getting better insulated and packed together. As we grow older, the white matter keeps changing, but scientists aren't quite sure about the details of how the white matter keeps changing. We still have lots to learn. In essence, when you are parenting you're not just growing a baby; you're essentially building a supercomputer from scratch, and you're doing it right from the first keystroke—conception.

Dr Catherine Lebel has been imaging the brains of children between the ages of two and eight years. Her work with MRI suggests that developmental changes in the brain at early ages have a significant impact on its functional properties in later life. Both brain structure and function can serve as indicators for future brain activity. These developmental trajectories are likely influenced by a combination of positive factors such as quality parent–child interactions and negative factors like drinking alcohol.

The Early Years: Building Blocks for Future Health and Success

The first years of a child's life are critical for their future well-being and success. The brain grows incredibly fast during this period, starting from before birth and continuing into childhood. Although our brains keep developing into adulthood, it's these early years that set the stage for all that comes after. However, it is never too late to make a change for our children or for ourselves—for our own sakes or for the sake of our parenting.

Brain development in a child is a complex interplay of various factors, extending beyond just genetic makeup. Key elements like proper nutrition starting from pregnancy, avoiding exposure to alcohol, drugs, and some medicines, toxins, and infections, and the child's interactions with people and their environment all play crucial roles. Consistent and nurturing care is fundamental for a child's healthy brain development. It's important to recognise that both positive and negative experiences during these formative years can leave a lasting imprint on the child. Therefore, parents and caregivers need ample support and the right resources to offer the nurturing environment that is essential for optimal brain development.

Children are naturally curious and eager to learn. They rely on their parents, family, and caregivers as their first teachers to help them develop the skills they'll need to lead independent, healthy, and successful lives. The quality of parent–child interactions is critical for healthy brain development. This has been given the term "serve and return" relationships, which we discuss later in more detail. Experiencing stress or trauma can have long-lasting negative effects on a child's brain, making it crucial for parents and caregivers to have the resources they need to provide stable, nurturing care. Tracking a child's development and achievement of milestones can help detect any issues early on, allowing for timely interventions if needed. Ensuring that all children have a healthy start in life is a public health priority.

The environment a child grows up in, including the emotional support they receive, plays a significant role in brain development. A safe, stress-free environment filled with opportunities for play and exploration is ideal. Parents can boost their child's brain development by engaging in conversation, play, and other nurturing activities. Effective learning occurs when parents take turns talking and playing with their children and build on their interests. Being attentive to a child's needs and providing sensitive care can protect their brain from stress. Reading, storytelling, and singing strengthen language and communication skills, setting them on a path for academic success. However, it's important to note that developing a healthy serve and return relationship with your children can also reduce risky behaviours. And remember, you only need to get this right part of the time! Remember that parenting is a journey, and no one expects perfection. No one expects 100%. Even 30–50% is often sufficient. Your efforts towards building healthy serve and return relationships will make a significant impact.

A Neuroscientist's Parental Awakening

Diving deep into the brain's intricacies, I often found myself lost in a land of neurons and synapses that often felt a world apart from the cosy bedtime stories, swings, and soccer practices with my kids. Yet a surprising revelation during my research bridged this gap, making me see the brain not just as an organ, but as the keeper of our unique potential, feelings, and experiences. Though we shared family moments and genetics, their minds were unfolding their own tales, dreaming their own dreams. This realisation reshaped my parenting. I started to view my children not merely as reflections of myself but as unique individuals. Their brains, like constantly updating supercomputers, process daily experiences, both joyful and challenging. This realisation helped me appreciate their authentic selves.

The lines blurred between being a scientist and a parent as both roles enriched each other. My deep dive into neuroscience and brain development, while aimed at understanding the brain and mental illness, inadvertently became a journey of understanding and connecting with my children's minds, hearts and life paths. My professional journey had given me one of the greatest gifts: a fresh lens through which to understand and see my children as the unique individuals they are. I started to rewrite my past parenting story.

Time for Reflection

Take a moment to think about your everyday parenting moments. Are there certain things you find yourself doing repeatedly? What usually sets off these reactions? Imagine you had a wise friend or "genie" to advise you on parenting—what might they say about you as a parent? Using this advice, consider how you could tweak the way you parent. Think about how understanding our brain's workings can help us become better parents. Finally, write down a promise to yourself about the kind of parent you want to be. Keep coming back to this promise and reflection as you navigate the ups and downs of parenting.

Chapter 3

KNOW YOUR ACTIONS

By Seeing Your Child's Silent Imitations

In Brief

You've probably heard of the phenomenon of yawns being "contagious" or of how you can almost feel someone else's joy or sorrow just by watching them experience it. These are instances of your mirror neurons at work. Mirror neurons give us an amazing peek into how children learn, especially from watching their parents. These special brain cells get activated when a child does something, but also when they see someone else do the same thing. For parents, this is big news! It means that because we spend so much time with our children, we have a big impact on their growth and learning. The discovery of mirror neurons marked a groundbreaking moment in the fields of neuroscience and psychology, providing new insights into language, social behaviour, empathy, and learning. These specialised cells were first identified in the early 1990s by a team of Italian researchers led by Giacomo Rizzolatti at the University of Parma. The researchers were conducting experiments on macaque monkeys when they noticed something intriguing: certain neurons in a monkey's brain fired not only when the monkey performed an action, such as grasping an object, but also when it observed another individual—be it another monkey or a human—performing the same action. This neural mirroring suggested a direct link between observation and action, opening up a new avenue for understanding a host of complex social behaviours and cognitive functions.

Research suggests that the mirror neuron system is most responsive to individuals with whom we share close, emotional connections. In the case of children, there is arguably no stronger or more influential bond than that with their parents. This means that the mirror neurons are likely to be most active and influential during interactions with parents, making the parents' behaviours, actions, and emotional responses especially potent as models for their children to emulate. We can leverage the power of mirror neurons as a guide for our children to follow us towards better behaviours, thought patterns, and more by changing *what* our children are mirroring—which means looking at our own behaviour. This is the fastest route to supporting your child's brain health.

Children Do as We Do: Mirror Neurons

We love our children, but our love is always tested when we are stressed or under pressure from our children's behaviour, our work, finances, and other outside and unforeseen forces. These are the precise lightbulb moments to reflect upon. Because it is in these moments that we pull out the only parenting skills we learned—from our own parents. Try to reflect on how you react to your children during your worst rather than best times. For example, what happens when you are overwhelmed with work and getting children to school? I know that I reacted poorly many more times than not as the level of stress and pressure rose in my career. Think of these responses as our stress reactions; they are how we learned to handle stressful situations as a child.

If we, as parents or carers, react abruptly to stressful situations or avoid conflict, it's likely our children will learn to react (not *respond*) to stressful events or avoid conflicts rather than face them and work through them. Consequently, our children end up facing the same issues we do. Yet we all want our children to have the best path forward and a bright future. So, how can we do this? As you probably already guessed, it all starts with us.

From the very beginning, our children, thanks to their mirror neurons, mimic our very behaviour. We might gasp, mouth and eyes wide, at them, and they might drop their mouth and open their eyes wide, doing the same. At the end of the day, we can't reproduce an action without seeing that action first, and this is where our mirror neurons play a huge role. We've all felt ourselves getting angry or stressed when someone close to us, such as our spouse or friend, is angry or stressed. These are your mirror neurons firing!

When a parent displays empathy, kindness, or patience, these behaviours are not just seen and registered by the child; they are neurologically mirrored. The child's mirror neurons activate in a way that helps them internalise these positive behaviours, facilitating the growth of similar neural pathways in their own brains. This can encourage the development of emotional intelligence, moral understanding, and social skills. The same can be said for skills like problem-solving, conflict resolution, and even more specific tasks like cooking or reading; children are keen observers and natural imitators, thanks to their mirror neuron system.

However, the potency of mirror neurons serves as a double-edged sword. Negative behaviours exhibited by parents—such as aggression, dishonesty, or unhealthy coping mechanisms—are also mirrored neurologically by the child. This could lead to the normalisation of such behaviours, laying a neural foundation that increases the likelihood of the child adopting these undesirable traits.

Therefore, understanding the power of mirror neurons adds a layer of responsibility to parenting. It underscores the concept that effective parenting is not just about the explicit lessons we try to teach our children but also about the implicit lessons we impart through our own behaviour. Parents need to be acutely aware that their actions, for better or worse, are being mirrored at a neurological level by their children. This makes the parent–child relationship an incredibly potent force in shaping who the child becomes, both emotionally and behaviourally. In essence, the mirror neurons magnify the adage that "children learn more from what you are than what you teach." Parents, then, wield an extraordinary power to shape the next generation, simply by being aware of their own actions and emotional responses.

CASE STUDY – Jack

Jack, a software engineer, always looked forward to weekends as it was his time to bond with his six-year-old son, Tim. They had a ritual of playing footy in the backyard every Saturday morning. However, Jack had a habit of getting easily frustrated when dropping the ball that would often result in a sigh, a roll of the eyes, or even an exasperated comment.

Tim, eager to make his dad proud, noticed these reactions keenly. His little mirror neurons fired away, capturing not just the mechanics of how to throw and catch the ball, but also how to react when things didn't go as planned. Soon enough, Tim started showing signs of frustration similar to his dad's. A missed catch would lead to a disappointed sigh, and a bad throw would make him throw his hands up in the air.

One weekend, Jack's sister Elena visited them. She couldn't help but notice the mirrored behaviours between Jack and Tim. After their game, Elena pulled Jack aside and spoke to him about the power of mirror neurons and the influence he had on Tim's developing emotional landscape.

The conversation was an eye-opener for Jack. He realised that his reactions were shaping Tim's understanding of how to cope with setbacks. The following weekend, Jack approached their game with a new mindset. When he missed a catch, he laughed it off and said, "No big deal. Let's try again!" When Tim fumbled, Jack encouraged him with a smile, saying, "You're getting better each time!"

As they continued to play, Jack noticed a shift in Tim's behaviour. His son was becoming more resilient and less prone to frustration. When Tim missed a catch, he picked up the ball, grinned, and said, "Let's go again, Dad!"

Building Serve and Return Relationships

Did you know that you can help build a child's brain—starting even before they can talk? A vital component of this nurturing connection is the practice of "serve and return" relationships, to use a term developed by the Center for the Developing Child (CDC) at Harvard University. Simple serve and return interactions between adults and young children help make strong connections in developing brains. And they're easy and fun to do. The child is on the "serve" side, and parents recognise cues from the child, such as questions, facial expressions, babbling, or reaching out to the parent. Parents/caregivers are the "return" side, which requires promptly setting aside distractions to respond with supportive words, encouragement, or affection.

Serve and return interactions are like a fun game that shapes your child's brain! When your child asks a question, babbles, gestures, or cries, and you respond with eye contact, words, or a hug, their brain forms strong connections for communication and social skills. When caregivers are sensitive and attentive to their child's needs, it fosters healthy development. On the other hand, when responsiveness is lacking, it can affect the child's physical, mental, and emotional health. So, let's strengthen these connections and support your child's lifelong learning and happiness together! The CDC have produced a series of YouTube videos that detail and showcase the beauty of this interactive process as adults and young children participate together.

The first three years of a child's life are undeniably crucial. However, it's important to note that they are *not* the sole determinants of a child's future. The brain retains a degree of plasticity into adolescence and even adulthood, allowing for ongoing development and learning. Therefore, opportunities to positively influence your child's development extend far beyond the toddler years. By maintaining a dynamic, responsive relationship with your child, you can continually enhance the quality of these serve and return interactions, fostering resilience, emotional intelligence, and cognitive skills across their lifespan.

The development of a child's brain architecture, which shapes their future academic performance, mental health, and interpersonal skills, relies heavily on the "environment of relationships" they experience. Unfortunately, many national policies overlook the significance of adult–child relationships for overall child well-being, missing the opportunity to nurture the hearts and minds of our future generation. Explore how you can integrate the following steps into your daily interactions with your child, fostering a loving and supportive environment for their growth and development.

> **Be Present and Attentive.** During the first three years of life, a baby's brain is hardwired to grab the attention of carers or parents. This helps them acquire the skills they will need (such as social, language, motor, and more). By receiving the attention they need in the early years, they build healthy neural networks and learn proper social, language, and motor skills from the get-go. Engage in one-on-one interactions with your child, giving them your undivided attention. Put away distractions like phones or other electronic devices when spending quality time together.

Observe and Respond. Pay close attention to your child's cues, gestures, and expressions. Respond promptly and warmly to them, whether they're smiling, crying, or otherwise demanding attention.

Listen Actively. Encourage open communication with your child. Listen actively to what they have to say, and show genuine interest in their thoughts and feelings.

Follow Their Lead. Allow your child to take the lead in play and conversation. Follow their interests and be willing to adapt your activities to match their preferences.

Be Affectionate. Express love and affection through physical touch, hugs, and positive affirmations. Physical contact helps build emotional connections.

Use Eye Contact. Maintain eye contact while interacting with your child. It helps them feel seen, heard, and valued.

Respond With Empathy. Acknowledge and validate your child's emotions. Respond with empathy and understanding, even if you don't always have a solution to their problems.

Stay Calm and Patient. Parenting can be challenging, but remaining calm and patient in stressful situations will foster a sense of security in your child. This takes a lot of practice; we will go into all the strategies in the chapter on stress and parenting.

Create Rituals. Establish daily routines and rituals, such as bedtime stories or family meals, to strengthen your bond and create shared memories.

Celebrate Achievements. Praise and celebrate your child's accomplishments, no matter how small. Positive reinforcement encourages them to explore and learn more.

Be Playful. Incorporate playfulness into your interactions. Play is a powerful way to engage with your child and promote their cognitive and emotional development.

Be Consistent. Consistency is essential in building trust and security. Try to maintain consistent responses and routines, as they provide stability for your child.

Between parent and child, physical displays of affection are key for good brain health and development. Children lacking affectionate parents often experience lower self-esteem and tend to feel isolated, aggressive, hostile, and antisocial. In contrast, *positive* interactions between parent and child that lead to increased child happiness and success include parental affection, such as hugs or a hand on the back, depending on the situation.

Studies have found that affectionate touch helps a baby's sense of touch, calmness, and ability to fight off illnesses to develop properly. It guides us to develop healthy social bonds and social behaviours and even verbal skills. When we hug or touch one another, our body releases feel-good chemicals like dopamine, indicating that physical touch between parent and child also contributes to the development of the reward system in the brain. And we will seek these feelings out in other ways if we haven't received them within our family unit.

A lack of feelings of warmth and closeness with our parents from a young age is linked with the diagnosis of life-threatening or chronic diseases later in life. One study showed that 91% of participants who reported not having a warm relationship with their parent had been diagnosed with diseases later in life, including alcoholism and coronary heart disease.

However, an accumulation of positive interactions with our parent leads to secure attachment in relationships later in life, as well as an increased ability to sufficiently regulate our emotions. And this all starts with our own self-regulation abilities. The more we work on ourselves, the more it positively influences our child's development.

Think of love as a skill, something that can be nurtured and grown within our minds. If we faced challenges in experiencing genuine love during our own childhood, it doesn't mean we're destined to carry those patterns forward with our kids. We can learn, adapt, and pour genuine affection into our family's life at any point. We have the power to access and nurture love at any stage in our lives, a concept we'll delve deeper into in a later chapter.

Positive parenting can even shape the amygdala (the fear centre) in the brain. Research shows that an increased frequency of positive maternal interactions changes the structure of the brain, hindering the development of mental disorders. Face-to-face interactions are key for brain development and, most importantly, healthy relationships, which research points to as being one of the most meaningful aspects in our lives. The more we prioritise these with our children and loved ones, the happier and healthier we might be!

Respond with Patience and Love

I remember a particular day, a day like any other in the rush of balancing work and family life. I had an important meeting scheduled and a strict deadline to meet. The clock was ticking, my anxiety was building, and it felt as though everything was conspiring to slow me down. My children did not want to go to school, and they were completely oblivious to the mounting tension in the air. I had asked them, perhaps more times than necessary, to get ready so I could drop them off at school on my way to work. But my pleas fell on deaf ears. Time was slipping away, and I could feel the heat of frustration rising. I snapped. My voice turned stern and louder than I intended. I said things in anger that I instantly regretted.

I saw their little faces fall. My heart ached. At that moment, my stress had taken control and turned into anger. Reflecting on instances like this with fresh understanding, I realised I was falling back on the only response I knew, one borne out of my own experience with stress-induced parenting. I wish I could say that these moments were a startling wake-up call. That they made me see how profoundly our personal stresses influence our interactions with our children. Unfortunately, I was not aware enough to make the changes needed at that time. It was not until later that I took responsibility to learn how to manage stress, to respond to my children with the patience and love they deserved, and to practise serve and return relationships. This was a challenging lesson to learn, but a crucial step in my evolution as a parent.

Inevitably, we cannot always pay super-close attention to our children, but one strategy is to bring your children alongside you, such as when making dinner or doing other daily tasks or going on a hike or when walking the dog. Deciding to strive to do the best you can and putting an effort in to nurture a serve and return relationship can fill your child's basic human need of safety, security, and stability. During times when it's not possible to fully immerse yourself in the interaction, acknowledging your child and letting them know you are unable to at that moment can, at least, offer some sense of care and safety. Even getting this right *part* of the time can go a long way. Remember, nobody is perfect!

Parents, remember that our kids look up to us. They're watching, learning, and absorbing, even when we might not realise it. By being aware, we can set the right example by living out the values we hope they'll embrace. And when we slip up? It's okay. Admitting our mistakes and showing them how to make things right teaches them that nobody's perfect. If ever communication gets tough and the road seems too rocky, it's alright to reach out for some guidance, like family counselling.

Remember, building a serve and return relationship is a journey that evolves over time but starts with one interaction, and even small efforts can make a significant impact on your child's development and well-being. Enjoy the process of connecting with your child and cherish the moments that shape your serve and return relationship. Sometimes, serve and return interactions may break down due to various factors. Adult caregivers might face significant stress from financial challenges, lack of social support, or chronic health issues, leading to difficulties in engaging with their children. Caregivers facing multiple problems are particularly at risk of providing inadequate care. It's essential to find support and a community of like-minded parents/carers to enable you to participate in serve and return interactions.

Establishing a Routine Check-In with Your Children

By designating a consistent yet flexible time each day or week for one-on-one conversations, you create an environment that encourages open and meaningful dialogue. One effective way to create a safe space for these conversations is by having meals together at a table, without the distraction of screens or devices. This setting naturally lends itself to more direct and meaningful communication, allowing you to focus solely on each other. It creates an atmosphere where everyone is more present, making it easier to delve into deeper topics and foster a sense of familial connection. Try to resist the impulse to criticise or offer unsolicited advice. Instead, strive to understand your children's perspective by actively listening and asking open-ended questions. Using "I" statements, such as "I feel concerned when you come home late", helps to express your feelings without sounding accusatory, thereby nurturing an empathetic environment.

Shared activities provide another pathway to support your relationship. Discover and schedule regular times for activities that both you and your children enjoy. Whether it's reading, walking, cooking, or working on a project, these shared experiences can be a fun and unobtrusive way to bond and encourage open lines of communication. Thanks to their mirror neurons, your actions will shape your children's brains and behaviour.

In a world that often feels like it's spinning too fast, children benefit from slowing down, tuning in, and being "seen"—because when it comes to parenting, the best tool we have is the quality of the relationship we build with our children.

Parenting Using Mirror Neurons in the Digital Age

So, we know that parent–child relationships and mirror neurons have a profound effect on brain development and behaviour. But what does this mean for parents in the digital age? Imagine scrolling through your social media feed and coming across a video of a friend laughing heartily. Without even realising it, you may find yourself smiling. This is your brain's mirror neuron system at work, translating observed experiences into your own emotional state. These mirror neurons don't differentiate between a real-life event and one viewed through a screen. Just as you can catch a mood from someone in your physical vicinity, you can also pick up emotional states through digital interactions. Ever noticed how a sad news article or an angry tweet can shift your mood?

Today's digital age is filled with moments that can trigger our mirror neurons, both in good and not-so-good ways. While tech can open our eyes to other people's experiences, helping us feel more empathy, it can also expose us to too much negativity, affecting our mood. By setting up a positive space and handling stress well, parents can boost their kids' brain adaptability. Being smart about tech and setting limits on phone and online time helps kids use the internet safely. Ensuring they get the right food, sleep, and activity is key to their healthy brain growth. And by promoting reading and a variety of experiences, we can help them think flexibly and solve problems more effectively. However, negative drivers, often mirrored from parents, can be detrimental. Unhealthy behaviours, like smartphone addiction or substance abuse, can adversely affect a child's development. Affluent neglect (where love is conditional on achievements) and inconsistent caregiving can disrupt a child's emotional stability. ACEs, like chronic stress or abuse, can negatively impact emotional regulation and brain development. Overexposure to screens can interfere with cognitive functioning, and a loss of familial and social connections can be emotionally damaging. Limited exposure to education can hinder cognitive abilities.

Time for Reflection

Pause and take a moment to consider your own actions. What behaviours do you notice your children picking up? Are they reflections of your own habits? How are your children when you are busy and/or stressed?

Chapter 4

FEELING CONNECTED

To Yourself, Your Child and Others

In Brief

One key factor that sits at the intersection of modern parenting and ancient wisdom is that it's the *quality* of our relationships that matters the most; it is not the number of people we know. It's the enduring power, protection, and safety we feel in healthy relationships. In a high-quality relationship, you have the unwavering assurance that at least one person has your back and is your steadfast support system, someone you can turn to no matter what life throws your way. It is not surprising that our ability to have quality relationships in our lives flows from the quality of our parent–child interactions. The steps are clear; if you have experienced high-quality interactions with parents and adults as a child, you will know how to give quality interactions to other people, and this will be returned to you.

How do we take the time to become curious about where our parenting skills came from? In a world governed by unrelenting expectations, the simplest yet most impactful gift we can give our children is the freedom to be themselves. This form of love isn't just about fulfilling immediate needs; it's about fostering a lifelong emotional and cognitive foundation that empowers our children to thrive in the face of life's complexities. This is the basis of resilience, an overused term that lacks the scientific understanding that every brain is different, both genetically and environmentally.

A Harvard study into aging was conducted from 1938 to 2004, led by the psychiatrist George Vaillant. While the initial emphasis wasn't on ideas like empathy or attachment, it soon became evident that the essence of healthy aging was deeply rooted in the strength of relationships. The researchers followed a large group of men until their death or later years. The goal was to determine what factors contributed to health, happiness and longevity. Every two years, these men were interviewed about their lives, including their emotional wellness. Researchers, much to their dismay, discovered that the quality of their relationships was the single most important factor. There was a strong correlation between happiness and the quality of close relationships, such as those with friends, family, spouses, and more. The researchers were hoping for something more quantitative like nutrition, exercise, career, status, power, or money. The conclusion researchers made was undeniably powerful. It showed, across a large span of time and with many different individuals, the importance of our relationships and the impact they have on our overall life and well-being. The quality of our relationships offers a major source of happiness, contributing to good brain health and mental wellness.

Parent–Child Quality Relationships Are Not Innate

My lightbulb moment happened while sitting in a lecture hall, listening to the latest research from Professor Michael Meaney, a renowned researcher in the field of the effect of parenting on brain development and the influence of previous generations. At that time, Meaney's work primarily centred around maternal care in rats and how it profoundly impacts the physical and emotional development of their offspring. He had mapped the brain architecture and was able to pinpoint that the parenting skills of maternal nurturing and care lay in the hippocampus, a part of the brain dedicated to learning and memory. Meaney's research sheds light on how the quality of parental care impacts everything from stress responses to cognitive abilities in offspring. The more nurturing the environment, the more resilient the offspring. Just as a rat mother's meticulous grooming of her pups impacts their stress reactivity, our daily interactions with our kids shape their emotional landscape in ways that are biologically embedded, shaping not just their minds but their futures.

This creates the skills that shape the quality of the relationships they enter as adults and determine who they are willing to let into their life. This has consequences for generations to come, especially regarding who they chose to have children with. They can enjoy a love that is nurtured, attentive, and yes, learned. And just like any other skill, it requires practice, patience, and a lifetime of learning. Meaney's research gave me a moment to reflect about my parenting skills. I listened as he discussed how a mother rat's care was sculpting the brains of her pups so that they learned how to go on and lick their own pups. He then presented data showing that a mother that did not lick her pups raised pups that did not lick *their* pups and so on through the generations.

My thoughts involuntarily drifted to my own parenting journey. My grandmother grew up in a small town in Western Queensland called Augathella, where everyone knew each other's name. When she was only seven years old her mother passed away. She found herself left to take care of seven brothers, an older sister, and a grieving father who lost himself in his work to cope with the pain. My grandmother stepped up, her childhood abruptly curtailed, as she started managing the household. She cooked, cleaned, and made sure her brothers got off to school. My grandmother would not have known that she never had the time to grieve, to be a child, to learn what motherly love felt like. She had become a caregiver before she could become a receiver of care. Then she was a mother herself. She gave my mum many more opportunities and an urban boarding school education, but the emotional language of motherhood—the quality interactions, hugs, the affirmations, the unconditional love—was foreign to my grandmother. She did her best, acting out of duty more than anything else.

Unconditional love is something my mother grew up without and was not taught. Then I came into the picture. She smothered me with love, was overly attentive, yet something was still missing. It was as though the missing emotional vocabulary had passed down through generations, like a book with pages torn out, leaving us to fill in the blanks in our own imperfect ways.

In that lecture hall something clicked. I realised that my parenting skills had been learned from previous generations. The clear message from the research being: your own parenting skills were lacking because your mother had not been cared for and passed this lack down to you. However, the brain is plastic and can learn a new skill—at any time, even later in life. It was then that I understood: to break the past parenting cycle, I had to write those missing pages myself, not just for me, but for the generations that would follow. And so, with knowledge, tools, patience, and the newfound wisdom that came from understanding my family's history, I began to learn how to create new mothering skills that focused on improving the quality of my parent–child interactions. I learned how to lick my pups, so to speak! Even though my children were teenagers by this time—so the interactions had to be age appropriate. I would argue that this can be the case for your adult children as well.

We often believe we love our children without bounds, that our affection is as limitless as the sky. Yet the journey of love is layered, complex, and demands a deeper exploration. The concept of unconditional love feels like it is innate, that we are born with it, as though it's woven into the fabric of our parental DNA. But, as groundbreaking research suggests, love—especially the unconditional variety—isn't purely instinctual; it's a skill, one that's developed and refined over time. This is a challenging thought, isn't it? It compels us to self-examine and ask: are we offering the sort of focused, unconditional love that not only comforts but also nurtures the very neural circuits of our children?

Happy and secure people make the most cooperative and responsive partners and the most sensitive and receptive parents. Inevitably, the opposite is true and can cause various problems in our lives, like jumping into low-quality relationships quickly or continually vying for the attention of others (and even pushing people away by doing so). After interviewing more than 150 people for the *Thriving Minds* podcast and in my life, I've come to conclude that we all seek love and attention—no matter who we are or where we came from. This is a biological need that all humans crave, young and old. If we don't receive this love in those early years, we will continue to seek out love and attention throughout our life, in good and bad ways.

One of the conversations that still lingers in my memory was with Gavin McCormack, who is a pioneering figure in the educational landscape. He has taught students from various corners of the world and has served in roles from Montessori teacher to school principal in Australia, global educational innovator, and co-founder of Upschool.co. The essence of his endeavours revolves around preparing the next generation with indispensable skills for the modern age. His groundbreaking approaches have been implemented from Australian classrooms to distant learning centres in the Himalayas, all converging on the idea of an education system where every life matters, optimism thrives, and nature is a treasured partner. He credits his tenacity and worldview to the daily words of encouragement his mother would gently share with him. The most impactful being her regular assurance of "I love you, unconditionally".

Phrases like "I believe in you 100%" and "I have your back no matter what" are not only free to say but also have a long-lasting impact that can span generations. However, if you've never heard these empowering words yourself, they may not even exist in your frame of reference, making it challenging to pass them on to your children. The words and the genuine sentiment behind them are essential for fostering healthy brain and emotional development at all ages. Despite this, society often doesn't guide parents to focus on this emotional groundwork when raising their children. It's worth taking a moment to question why such vital aspects are commonly overlooked.

Unconditional love is a ***special kind of love*** and is a key driver and protection of our brain health across the lifespan and the generations. This is a confrontational notion, and it means accepting your children for who they really are, not for what you envision they could or should be. This is hard as a parent to absorb or to be aware of because we are the products of our culture, and our families, and our parents' expectations.

Unconditional love is the key to happiness, health, and even our survival as a species. This is accessible by everyone, but it is the one ingredient many people are missing from their lives. We often hear, "It takes a village to raise a child." But on the other side of this equation ... if a child doesn't receive the love, attention and support they need early in life, it can actually burn the whole village down.

What Is Unconditional Love?

When parents accept, love, and show affection to themselves and their child, even when they make mistakes or fall short of expectations, this is unconditional love. In other words, it is a form of love with no strings attached. Parents who love their children unconditionally love them for who they are, no matter what. This is what we aim for, and in our minds, it feels like it's what is happening. But here are a few scenarios to think about that often test the boundaries of our unconditional love.

Imagine your child, George, has always been excellent at maths and science. You've always envisioned a future for him as a doctor, engineer, or in some other profession. You even have family members in those fields who could offer guidance and mentorship. But one day, George comes to you and announces he wants to be an artist. He's passionate about painting and wants to dedicate his life to it. While family and friends might express disappointment or concern, unconditional love would mean supporting him in pursuing his passion, even if it goes against your expectations or societal norms.

Raised in a devoutly religious family, your son Dillon has been an active member of your community church since childhood. As he grows older, he starts questioning his faith and eventually identifies as an atheist. While this could potentially cause rifts in the family dynamics and be a topic of contention, showing unconditional love means accepting his right to his own belief system, even if it diverges from your own.

Sasha, your youngest, comes out to you as gay. The family has had a conservative upbringing, and you know that this revelation might not sit well with some of the older family members. However, unconditional love entails supporting Sasha's identity, being her advocate, and educating both yourself and others in the family to create an environment where she can be her authentic self.

Your daughter Jasmine has grown up in a community where everyone you know has followed a similar path—university, career, marriage, and then children. Jasmine, however, tells you she wants to travel the world, engaging in humanitarian work without settling down in one place. While the conservative circle around you might see this as a phase or a waste of potential, embracing Jasmine's choices with unconditional love means respecting her individual path and encouraging her to find happiness her own way.

This is a small subset of the infinite possibilities we encounter across our lifetime, not only from parents, but educators, friends, employers and society in general. Parents are the epicentre for children, and this is true for the lifespan. Unconditional love is a superpower, and its effect on brain health is lifelong and lasts for generations. It is the shield to protect against life's adversities and complexities. It is the elixir of our human experience, and it is the one thing that is free to give and that each of us can give. There are many reasons that this is hard for people to lean into because of early life experiences. We will learn how to unlock it and learn about neuroplasticity throughout the rest of the book.

We think that all parents love their children—like it's a given. However, at the first sign of a misstep or bad behaviour, we have a haunting pervading sense that perhaps our child is not going to measure up. This can happen as early as preschool for some children, who are shuffled from violin to piano lessons, absorbing multiple languages before they've even mastered their native tongue. But the most important gift any person can receive is our unconditional love.

Time for Reflection

What might be the first step you take to enhance the quality of your relationships with your children, regardless of their age?

Chapter 5

BEING TECH AND SEX SAVVY

Having Digital Literacy Skills

In Brief

Recently, I was sitting in a circle at a teacher's staff meeting, and one of the teachers shared a story about asking her class what they wanted to be when they grew up. One of her students, Samantha, popped up her hand and said, "An iPhone!" The teacher was taken aback and thought, at first, perhaps Samantha hadn't heard the question. The teacher asked, "Why do you want to be an iPhone, Samantha?" She answered, "Because my mum and dad love their iPhone more than anything." A bit shocking, right?

We all know technology has changed our lives in many ways. But what we might not realise is how much it has changed our relationships, our brains, and our kids. When a child thinks they're less important than a phone, we know something needs to change. So who is actually looking out for our kids to keep them safe in a digitally immersive world?

The difficulty is genuine for parents trying to manage their children's screen time and technology usage. Gone are the days when the biggest concern was how many hours a child spent watching television; now parents are faced with an intricate web of challenges—from smartphones and social media to online gaming and virtual classrooms. This ever-present digital landscape not only complicates the task of setting boundaries but also raises pressing questions about technology's impact on children's mental health. Balancing technology's undeniable benefits with its potential downsides has become a high-stakes juggling act, leaving many parents feeling overwhelmed, conflicted, and in search of practical guidance.

> **CASE STUDY – Meera**
>
> In the dim glow of a late-night living room, Meera stared at the array of devices scattered across the coffee table—tablets, smartphones, and a gaming console remote. She had spent the evening negotiating screen time with her two children, navigating the murky waters of parental controls, and grappling with the unsettling feeling that technology was becoming the third parent in her home. As she looked at the unread emails on her own phone and notifications from her children's school about another online safety seminar, she felt an overwhelming sense of defeat.
>
> The simple decisions of parenting seemed like a thing of the past. Now Meera found herself wrestling with a constantly changing digital world that felt like it was taking over her family's life. Surrounded by the soft buzz of the devices and glowing screens, Meera let out a scream and thought, *I have no clue what to do next.*

It's important to extend compassion and understanding to all parents navigating the intricate maze of raising children in the digital age. We're all learning on the fly, often juggling multiple roles and responsibilities, while also contending with a rapidly evolving technological landscape that many of us did not grow up with. The challenges are real, but so are the opportunities for growth and adaptation.

If the digital age has taught us anything, it's that the only constant is change. As parents, this means our approaches must be equally fluid and dynamic. While the traditional parenting styles provide a foundational understanding of child development, they are not one-size-fits-all solutions in today's complex world. And that's okay. Our wisdom and life experiences are invaluable assets that can guide us in blending the best of the old with the new.

You don't have to overhaul your entire parenting approach to make room for the digital age. Sometimes, all it takes is a few open-ended questions to ignite meaningful conversations with your children. Questions like "What was the most interesting thing you saw online today?" or "How do you feel when you spend a lot of time on your device?" can serve as simple but effective entry points into your child's digital world.

The first step in adapting to this new reality is recognising the limitations of traditional parenting styles, which were formulated in a pre-digital age. While these styles offer frameworks for understanding human behaviour, they often fall short in addressing the unique challenges posed by the digital landscape—be it cyberbullying, screen addiction, grooming or the complex social dynamics of online interactions.

What Is Your Current Parenting Style?

Research indicates that our parenting styles are heavily influenced by how we were parented. While traditional literature identifies four main parenting styles, each with its unique impact on child behaviour and development, these models need to be adapted for the complexities of the digital age. At the end of this section, I introduce the concept of "Socratic Parenting", which offers a fresh perspective, equipping parents with the tools to engage in meaningful dialogue with their children about their digital lives, thereby fostering critical thinking and responsible digital citizenship. This is a new approach designed to help parents navigate the challenges of raising children in a digital-saturated environment.

Authoritative Parenting (Democratic)

About 50% of parents employ the authoritative parenting style, where parents set clear rules and expectations but also allow for some flexibility. They engage in a serve and return relationship with their children, valuing open communication and fostering autonomy through natural consequences. Research shows that children raised in this manner often exhibit a range of positive outcomes, such as improved happiness, increased self-confidence, and lower risk of mental disorders. This approach may not fully address the unique challenges presented by the digital age.

The digital age operates on different rules, where the consequences of actions are not always immediately apparent. For example, a child may not understand the long-term implications of oversharing personal information online, and the "natural consequence" of this action could be severe, such as identity theft or cyberbullying. Also, the authoritative model may not fully equip parents to discuss the ethical considerations and addictive nature of digital platforms, which often require a more nuanced approach.

Authoritarian Parenting (Disciplinarian)

This is the second most common type of parenting style. In this model, rules are rigidly enforced with little consideration for the child's emotional needs or perspectives, often justified by the "because I said so" rationale. Communication is largely one-directional, from parent to child, stifling any attempts at dialogue. Research has shown that children raised in authoritarian households are more likely to experience a range of negative outcomes, including lower self-confidence, academic struggles, and a higher risk of substance abuse. In the context of the digital age, these pitfalls become even more concerning. Children raised in authoritarian homes may lack the critical thinking skills needed to navigate the ethical and safety challenges posed by the online world, from cyberbullying to privacy concerns. Their inability to openly communicate with their parents could lead to a lack of supervision or guidance in their digital lives. Moreover, the disciplinary focus of authoritarian parenting can deter children from discussing their online experiences openly, out of fear of punishment. This can result in children hiding their online activities, making them more vulnerable to online risks.

Permissive Parenting (Indulgent)

Some parents adopt a permissive or "passive" parenting approach, in which parents are highly responsive and make few demands. While this style often results in open communication and potentially secure attachment, it leaves much to be desired when preparing children for the challenges of the digital age. In this environment, children are given considerable freedom and minimal guidance, allowing them to navigate the digital world largely unsupervised. Rules around screen time, online behaviour, and digital safety are either lax or non-existent.

The permissive style's shortcomings become glaringly evident as children face the myriad challenges of online life—from cyberbullying to exposure to inappropriate content. The lack of structured guidance can make it difficult for these children to develop the self-discipline and critical thinking skills necessary to navigate complex online ecosystems responsibly. The tendency for permissive parents to avoid conflict for the sake of keeping their children happy may also prevent meaningful discussions about digital ethics, safety, and the consequences of online actions.

Moreover, the "friend over parent" dynamic characteristic of permissive parenting can blur the lines of authority and accountability, making it challenging to enforce any digital guidelines that may be introduced later. Children raised in such environments may struggle with self-control and may exhibit egocentric tendencies, which can manifest as risky or thoughtless online behaviour.

Neglectful Parenting (Uninvolved)

Neglectful parenting, although less common, presents specific risks in the digital age, particularly around issues like grooming and sextortion. This is where parents are not responsive and make no demands. This often stems from parents' emotional unavailability due to work, social commitments, or their own unresolved mental health issues. The result is a lack of emotional support and guidance for children, which can lead to insecure–avoidant attachment patterns affecting relationships throughout their lives.

In the context of our increasingly connected world, the implications of neglectful parenting take on new dimensions. Children raised in such environments are often left to navigate the digital landscape with little to no oversight, making them especially vulnerable to online predators. The lack of emotional connection and open dialogue means that children are less likely to report uncomfortable or dangerous online interactions to their neglectful parents. This absence of a trusted adult figure to confide in opens the door to risks like online grooming and sextortion, where a child may be manipulated into sharing personal or explicit content.

Moreover, these children are also at a higher risk of falling into digital pitfalls, such as cyberbullying or addiction to online games or social media, as they might seek the emotional support online that they do not receive at home.

Therefore, it's crucial to recognise the signs of neglectful parenting and understand its unique implications in the digital age. Intervention and education are vital for the emotional well-being of the child and for their digital safety. Neglectful parenting not only deprives children of emotional security but also leaves them exposed to the darker elements of the digital world.

Affluent Neglect (Parenting When You "Have It All")

The term "affluence" often evokes images of financial abundance and material comfort. However, in the context of parenting in the tech-savvy 21st century, affluence can paradoxically create an environment ripe for neglect. In a digital age where smartphones, tablets, and high-speed internet provide constant connectivity, affluent families face unique challenges that aren't always obvious.

Though countries like Australia have seen remarkable economic growth, the "income paradox" shows that a rise in affluence doesn't necessarily translate to long-term happiness or emotional well-being. In affluent families, parents may find themselves ensnared in a work-centric lifestyle, sacrificing crucial family time and emotional engagement. This creates a gap that is often filled by digital devices, as parents may use technology as a substitute for personal interaction, assuming that financial privilege can compensate for emotional absence.

Affluent neglect takes on a new dimension in the era of screen addiction and online risks. While parents might invest in state-of-the-art educational software, private tutors, or enriching extracurricular online programs, they might overlook the need for emotional availability and conversations about digital behaviour and ethics. This form of neglect is often masked by affluence, leading to a form of "digital isolation" where children may turn to online platforms for the emotional support they're not receiving at home. The consequences can range from cyberbullying to online predatory risks and even to a distorted sense of self-worth and validation-seeking through social media likes and follows.

Moreover, studies indicate that children in affluent settings experience heightened levels of anxiety, about 20–30% more than their less affluent counterparts. They're also more vulnerable to substance abuse, which now extends to digital addictions, whether it be to social media, gaming, gambling or problematic internet use. This susceptibility often arises from the high-performance expectations set by successful parents and is exacerbated by the constant social comparison facilitated by digital platforms. The challenge isn't just to equip children with the material resources they need to succeed but also to provide the emotional and ethical guidance they require to navigate the digital landscape safely and responsibly. Affluence may offer many advantages, but in the digital age, emotional availability and parenting are resources that no amount of money can replace.

Helicopter and Snowplough Parenting (Overprotection and Micromanagement)

New terminologies in parenting like "helicopter parenting" and "snowplough parenting" have emerged to describe highly involved parental styles. Helicopter parents are notorious for their overprotective nature and a tendency to hover around their children, much like a helicopter. Snowplough parents go a step further by not only micromanaging but actively clearing obstacles from their children's paths.

While these styles may seem advantageous in a digital age full of online risks and challenges, they come with their own set of drawbacks. From a neuroscientific perspective, these styles can hinder a child's development of essential life skills like resilience, stress management, and a growth mindset. Children brought up in such environments may miss out on learning how to cope with the natural failures, discomforts, and stresses that are part and parcel of growing up.

In the context of technology, this over-involvement can manifest as excessive monitoring of screen time, social media activity, and even homework done on digital platforms. While it's important to be aware of your child's online behaviour, over-monitoring can prevent them from learning essential digital literacy skills, such as discerning credible from non-credible information, understanding the importance of privacy settings, and developing the emotional intelligence required to navigate social dynamics online.

Moreover, children raised by helicopter or snowplough parents might not learn how to set their own boundaries in the digital world. This opens the door to potential issues such as digital addiction, poor time management, and even vulnerability to online predators who prey on those who haven't learned to set and maintain online boundaries.

Combating the urge to become a helicopter or snowplough parent, particularly in this digital era, often starts with self-awareness. Questions to ponder include "What type of parent am I currently?" and "What type of parent do I aspire to be?" Reflecting on your own upbringing can provide invaluable insights into your current parenting style and guide you towards creating a balanced, nurturing environment where your children can thrive, both online and off.

Enter Socratic Parenting

Socratic parenting is a modern approach that blends the wisdom of age-old philosophies with the challenges of the digital age. This style encourages open conversations, critical thinking, and ethical reasoning, equipping children with the skills they need to navigate the digital world responsibly. Unlike traditional styles, Socratic parenting goes beyond rule-setting, to foster a sense of inquiry and understanding, which is vital in an age where children are exposed to a barrage of information and influences online. The goal isn't to win the argument, but to really think deeply about the topic and understand it better. It's like shining a flashlight on an idea from different angles to see it more clearly. This kind of discussion encourages everyone involved to think more critically and come to their own conclusions.

Socrates, the ancient Greek philosopher, employed this method to engage his students in deep questioning to help them arrive at a truth. Socratic parenting can be a highly effective method for fostering critical thought and strong parent–child relationships, but it's essential to adapt the approach to suit the age, maturity, and individual needs of each child.

How to Create Open-Ended Conversations

Drawing from the Socratic method, parents can learn how to use open-ended questions to encourage children to think deeply about a subject. Instead of providing immediate answers, the parent facilitates a conversation that enables children to arrive at their own conclusions. This approach not only empowers children but also makes the learning experience more engaging and rooted in critical thinking. This method helps children to internalise the reasons behind guidelines and rules, making it more likely they'll follow them. It encourages kids to analyse situations critically, which is a crucial skill in navigating the digital world safely. It is another way to strengthen the parent–child relationship; the Socratic questioning, unlike the dictatorial form of communication, fosters an atmosphere of mutual respect and understanding. By involving them in the conversation, children feel a sense of ownership over their actions, making them more responsible digital citizens.

Learning how to initiate and sustain open-ended conversations is an invaluable skill for parents aiming to deepen their relationship with their children. Open-ended conversations lead to high-quality parent–child interactions as we gain knowledge about the children, and can be particularly useful in heated moments, because they encourage conversations that help to diffuse tension. If we have not learned these techniques, it is never too late to start. The good news is that our brains are malleable; we can rewire neural connections through consistent effort and practice. This means that even if we've inherited certain parenting behaviours, we're not stuck with them. To truly prepare our children for the world they will inherit, we need to update our parenting "operating systems".

Remember, parenting is a journey, not a destination. As we adapt to the needs of a new generation, we're rewriting the analogue codes that governed our own upbringing. By doing so, we're not just preparing our children for the world they will inherit; we're joining them in shaping it. With a bit of wisdom, conversations, and a willingness to evolve, we can successfully guide our families through the challenges and opportunities that the digital age presents.

In an era where banning technology for children is neither practical nor entirely beneficial, parents find themselves at a crossroads. The complexities of the digital age demand that we adapt and evolve our parenting techniques to meet contemporary challenges. Socratic parenting emerges as a crucial skill set for the digital age, offering a pathway to engage our children in critical thinking and responsible decision-making. But beyond techniques and strategies, the heart of successful parenting in this digital age lies in the quality of the relationships we build with our children. By fostering open dialogue, mutual respect, and trust, we can equip our children with the tools they need to navigate the digital world and also create a familial environment where they feel secure and understood. The ultimate goal is not just to manage screen time or online activities but to cultivate relationships so robust that our children know and trust us implicitly—and, more importantly, feel known and trusted in return.

Time for Reflection

Considering the evolving nature of parenting styles, especially in today's digital age, how might you adapt or introduce a new approach that not only strengthens your bond but also prepares your child for the world they are growing up in?

PART 2

Barriers to Parenting

Chapter 6

LIVING IN THE 21st CENTURY

Parenting Children in a Stressful World

In Brief

Every parent begins their journey with genuine intentions to be an effective parent by offering the best for their child, hoping to guide them on a path illuminated by love, comprehension, and guidance. Yet the modern world introduces hurdles that makes it harder to be the parent that we would like to be. Among these hurdles, everyday stress stands as the most formidable barrier to nurturing healthy children. In today's fast-paced world, many things demand our attention, often reducing the quality time we spend truly seeing and understanding our child. Furthermore, the essence of parenting, by nature, is imbued with stress. Concerns for our children's well-being, self-doubt about our parenting choices, and anxieties about the broader world—from environmental crises like climate change to economic downturns—consistently weigh on our minds.

The relentless hustle of modern life, be it from work pressures, financial worries, social commitments, children's engagements, or the omnipresent pull of digital media, constantly strains our mental bandwidth. This constant tug-of-war for our attention invariably breeds stress, making the beautiful journey of parenting simultaneously challenging. Our parents faced sporadic and intense stressors.

In contrast, today's society sees many people grappling with the unyielding burden of chronic daily stress. This shift has contributed to a surge in mental health disorders over recent years. Even before the COVID-19 pandemic, stress had been recognised by the World Health Organization (WHO) as the defining health crisis of the 21st century. In our fast-paced world, stress is a constant. It's not just about us; it also affects how we parent our children. Since stress isn't going away, we need simple daily ways to handle it. By managing our stress, we not only help ourselves but also set a positive example for our kids, showing them how to cope in tough times. After all, being a good parent means taking care of ourselves too.

How Stress Affects the Brain: Our Smoke Detector

For our ancestors, the physiological response to stress was helpful and necessary for survival. If a bear or tiger leaped out in front of them, these hormones gave them the quick energy they needed to fight or flee. Yet this doesn't translate in the modern era. Today, we don't usually have bears or tigers to flee from; our stress may manifest as yelling, quick and negative reactions, or other behaviours that we don't fully think through before implementing. And today, we have different stressors. Whether it's the relentless pings from our devices, the pressures of juggling work and family, or the ever-present concerns about our children's well-being, our internal alarms—our amygdalas—are constantly ringing. When this happens, the amygdala sends signals to various parts of the brain and body, activating what's known as the "fight or flight" response. This heightened state of alert can not only impact our own health by elevating stress hormones and increasing inflammation but also deeply influence our relationships with our children.

The idea of calling this part of the brain the "smoke detector" was put forward by Dr Randolph Nesse, founder of evolutionary psychiatry. A "real" smoke detector frequently goes off, blaring false alarms, such as when we forget to turn the fan on when cooking. In the chaos of modern life, our stress response often operates like a hyper-sensitive smoke detector, going off at the slightest whiff of trouble. This smoke detector in neuroscientific terms is the amygdala, a small almond-shaped cluster of nuclei in the brain responsible for emotional processing, including fear and stress reactions. The amygdala is intricately connected to other brain regions and the body, influencing everything from heart rate to attention.

This area of the brain will prioritise stress over happiness, *always*. In other words, when stressed, it takes over. Think about how you feel when you're stressed; it's hard to stay in control. For the most part, we just want to get rid of it or let it out. This is due to many reasons, particularly how our brains have been wired by stress over many generations, which leads to rising cortisol and adrenaline in the body. Because each of us has experienced different amounts of stress, this means we react differently to rising cortisol, adrenaline, and other neurotransmitters.

Just as a smoke detector that goes off too frequently can become a source of annoyance rather than a safety feature, our constant state of stress, driven by an overactive amygdala, can become disruptive. It can cloud our judgement and reactions in crucial parenting moments. When we're in this state, the prefrontal cortex—the brain's centre for rational thinking—often gets bypassed, making it harder for us to make thoughtful decisions.

Understanding the neuroscience behind our stress reactions can help us manage them better, not just for *our* well-being but also for the emotional health of our children. Learning to regulate our smoke detector through understanding neuroplasticity and knowing how to pause the smoke detector is the first step to creating a calmer, more nurturing environment for ourselves and our families. While we can't control the traumas we've experienced, we can control how we respond to them with support from caring and well-trained professionals and the actions we take every day to retrain the sensitivity of the brain's smoke detector. This involves understanding our origins and finding purpose in our lives.

While it's good to be informed about what's going on in the world, excessive consumption of social media and the news and the increasing use of digital devices contribute to increased stress in our lives. We are now more likely to compare ourselves to others, even though what others broadcast on social media is likely only their highlight reel. And here's the other thing: bad news sells. This means we are more likely to consume negative news or recent events than positive, leaving us feeling depleted and overwhelmed.

As parents, it's a lot. Yet understanding how stress works in the brain can help us overcome it and become better individuals and parents. And don't expect overnight success, but rather focus on the process of learning and improving every day. Laying down that one brick at a time eventually leads to greater change and major growth. This chapter aims to help you "pause the smoke detector", allowing you to turn knee-jerk stress reactions into thoughtful responses. By understanding the neuroscience behind stress and its effects on our brain, you'll gain valuable insights into how you can manage your stress more effectively. More importantly, you'll learn how to impart these lessons to your children, equipping them with the emotional and cognitive tools they need to navigate their own stressors—both online and offline.

Is All Stress Bad?

All this might make you wonder if *all* stress is bad. The quick answer is no. Neuroscientist Tracy Denis-Tiwary's research goes into detail about how stress and anxiety can make us more resilient. Many of us, when we experience increased stress or anxiety, immediately think it's bad and that we need professional help and treatment. Yet in many cases (not to deter anyone from seeking the help they need), what we need to focus on is building skills to cope with and distinguish between difficult and negative feelings. We don't always need to be diagnosed with an anxiety disorder or disease. Anxiety is a normal part of being human.

In many ways, this relates back to evolution and understanding why we feel stressed and what's going on in the brain when that happens (as described above). On the other hand, anxiety revolves around fear of the future. But it also gives us hope. If we feel anxiety, we may work harder to achieve positive outcomes; this isn't inherently bad, nor is it necessarily a bug or malfunction occurring in the brain, as many of us may think.

We can leverage stress and anxiety for good. They can help us stay motivated and fuel our drive to achieve what we want for ourselves and our kids. And, most importantly, we can learn from anxiety and stress and then, after leveraging them, simply teach ourselves to let them go. In turn, this allows us to return to the present moment and use stress as a tool, as opposed to an obstacle or hurdle we must overcome.

This perspective on anxiety is known as the "virtuous cycle", which includes the three L's:

- Listen
- Leverage
- Let go.

For example, maybe we start panicking that our child didn't do well on their report card and that they aren't on track to getting into a university or a trade or work. Instead of reacting, let's take a moment and listen to these thoughts. Then, let's leverage these thoughts by playing out the benefits of being the parent of an enterprising young adult who is pursuing interests and talents that suit them. Now, it's time to let go of those anxious thoughts with the knowledge that you've stepped back and viewed this situation from a wider lens; this is exactly how we can use anxiety to our advantage!

How Stress Impacts Our Cognitive Function, Wellness, and Brain Health

Stress can further provide us with a useful tool for understanding ourselves, fuelling personal growth, and enhancing our overall health. It doesn't have to hinder or block us from being who we want to be, especially when it comes to parenting our children. When we retrain the brain with small but regular practices, we can respond how we want, as opposed to simply reacting.

As previously mentioned, chronic and persistent stress can end up taking its toll on not just our minds but also our bodies and our entire lives. I remember this incident almost as if it was yesterday: I knew I was gaining weight; my clothes were fitting a little more snugly. I didn't feel particularly comfortable in anything I wore. That morning I stepped on the scale, and I cried. At 44 years old, I was the heaviest I'd ever been.

Looking back now, it's no surprise *why* I gained weight. On top of a stressful career, raising two children, staying connected with family and friends, and caring for my family, I had recently lost my beautiful sister. I immersed myself further in my work, working most days *up to ten hours*. I didn't have time for exercise or preparing healthy meals. I wasn't doing the basics for myself. I was merely chugging along, fuelled by stress. And my stressed-out brain was telling me to eat and drink more.

Later on down the line, I began to understand the neuroscientific explanations behind this mechanism. My research showed how once you get on that sugar train, it's hard to get off. Sugar *is* addictive, and stress leads the way with this addiction by driving us to eat foods high in fat and sugar. I wasn't immune to this. And, like everyone else, I was also partially coping with increasing stress and striving to feel better via alcohol and comfort foods.

The moral of this story? I was being controlled by my stressed brain rather than the other way around. Yet the discovery I made was that I could teach my brain how to respond to stress better. I *could* change the unconscious habits that were driving my weight gain. And yes, this came down to learning how to train that smoke detector.

The brain's "smoke detector" is among its most ancient components, with roots going back around 250 million years. This is where emotions such as fear, stress, and pain arise, often due to generational trauma and adverse childhood experiences (ACEs). Ultimately, the brain's primary role is ensuring our *survival*, not necessarily our *happiness*. This might explain why many individuals grapple with mental health challenges and achieving life satisfaction. When under stress, the brain prioritises immediate survival, even if the perceived threat isn't genuinely life-threatening.

If you've seen the viral video of the cat and the cucumber, you'll have a very clear visual representation in your mind as to how exactly the stress response works. The video begins with a cat busily eating its food while its owner places a cucumber beside it, out of its field of vision. When the cat turns around, it's startled by the cucumber, jumping into the air with fear. Perhaps it automatically assumes the cucumber is a snake or predator of some kind. Either way, the stress response happens in a split second. When the smoke detector receives a stimulus, it processes it in mere milliseconds. Within this time, cortisol and adrenaline get pumped out. With strong neural connections between the amygdala, heart, lungs, stomach, and legs, our heart rate quickens, our breathing becomes shorter and shallower, our palms get sweaty, and all of our muscles become tense. We run away or freeze or fight. Just like the cat in the video, *we become stressed!*

With ongoing stress, our digestion also takes a hit. This is why we might experience symptoms like diarrhoea, constipation, heartburn, stomach cramping, or general indigestion. The brain-gut axis plays a major role in connecting the brain and stomach, meaning how we *feel* can impact our digestion. That "gut feeling" isn't just something we hear in stories; it's quite real.

These automatic and innate responses to stress are why it can feel impossible to override our initial reactive responses (even though it is *entirely* possible to override them with the right tools, brain training, and practices). The main reason this is difficult is that the smoke detector is made to identify threats or dangers. The result is that we tend to remember negative experiences more than positive ones. Remembering these negative incidents or events can help us recognise potential threats to our survival and, hopefully, avoid them.

In fact, the cerebellum area of the brain is very good at holding onto traumatic memories; then, when recalling these traumatic memories, it can feel as though the trauma is happening all over again, causing individuals almost to relive those emotional reactions. Our brain is working hard to ensure we have what we need to fight or run away from threats. For our ancestors, these threats were brief, and once the threat disappeared, so did their stress. For us, our brain perceives an abundance of entities in our lives, such as the news and social media, as stressful. Many things are tugging at our attention—more than ever before. And because of this, many of us end up *chronically* stressed.

So, What Exactly Happens in the Brain with Chronic Stress?

As previously mentioned, chronic stress drove me to eat an increasing amount of unhealthy food and drinks, resulting in significant weight gain. Over time, the stress hormones can further significantly harm the brain and interfere with its communication pathways. *But the brain is very smart.*

It eventually balances these harmful chemicals with feel-good chemicals. However, we need a way to stimulate these chemicals in the reward centre of the brain. So, what does the brain do? It drives us to eat sugar or fatty and greasy food or take that sip of alcohol or have a hit (or two) of that cigarette or smartphones and devices. This can help explain why, when we are emotionally distraught, we turn to comfort foods like ice cream or mac and cheese, or less-than-healthy behaviours.

Now, here's the good news: we aren't tied to these actions as the *only* ways to activate the reward centre of the brain. We can also stimulate those feel-good chemicals through healthy activities, like exercise, socialising, and playing with our children. At the same time, if we've been turning to unhealthy ways to manage stress (which is different in different people), this change may prove harder to do (but, again, not impossible!).

Yet, how we react to stress each and every time teaches the brain how to handle it in the future. This is the power of neuroplasticity! Considering that most stressful events in our lives aren't actually life threatening, it's highly beneficial to strive to create space between the stress stimulus and our response, such as by pausing and responding how we would like to as opposed to allowing the smoke detector to guide us. This means we need to learn to properly handle stress when it inevitably arises—and (not to sound like a broken record here) we can do this by training our brains as we would our muscles.

We can actually improve our minds by training our brains. So we will explore how we can fine-tune our internal alarm systems, not just for our sake, but for the sake of our most precious relationships, in Part 3. After all, the quality of the time and attention we give to our children begins with our ability to be present, and that starts with learning how to pause the smoke detector.

Time for Reflection

Recall a recent stressful event. Before reacting to the event itself, try to identify the underlying cause of your stress. Was it a direct result of the event, or was it influenced by past experiences? Now envision your brain's "smoke detector". How often does it go off? Are there specific triggers that make it more sensitive?

Chapter 7

ECHOES FROM OUR PAST

How Our Childhood Shapes Our Parenting

In Brief

Our childhood memories and experiences play a big part in how we raise our kids. Sometimes, we might parent in ways shaped by our past without even realising it. By understanding these influences, we can make better choices and avoid repeating old patterns. Let's explore how our early years impact our journey as parents. Modern DNA tests show we inherit traits from our ancestors. This makes us wonder: are our reactions based on our own experiences, or even those of our parents or grandparents? Knowing this helps us be kinder to ourselves, understanding that our responses might come from old family patterns. With this knowledge, we can better understand our parenting and create a loving space for our kids. Building on this, how we parent has lasting effects, especially on our children's mental health. Just as we might carry traits or reactions from our ancestors, our parenting choices can influence our kids' mental health and well-being for years to come.

In 1989, I was sitting at my desk writing up a thesis about why people buy cough and cold medications, sitting next to an enormous computer that went from floor to ceiling. The phone rang, Mum on the other end. "It is your sister, something is wrong, come quickly."

Francesca had arrived home on a flight from Sydney, where she had been living. She was different, in a way that was hard to pinpoint. I was about to complete a pharmacy degree and thought I had it planned out. I was following in Dad's footsteps; the future was bright and clear. The next morning, we went to the emergency department, and within a couple of hours, Francesca was sent to Rosemount, which in the 1980s was used as a psychiatric facility for the nearby Royal Brisbane Hospital. It is hard to imagine now, but during this period the operating theatre was used to administer electroconvulsive therapy (ECT or "electric shock treatment").

Entering the lock-up ward for the first time was like going to another planet. We arrived at a devastating situation. Sitting upright, stiff as a board, staring into space, in the corner of a dark smelly room, in the middle of a psychiatric lock-up ward, was Francesca. Her empty eyes staring straight ahead as I helplessly watched the haloperidol-filled needle being inserted into her arm. We were in an old demountable unit, surrounded by lost souls wandering around half naked, in various states of screaming and crying. This was far from the fantasy of our life—a fantasy that ended after entering the front door of this unit.

Many doctors were seeing Francesca only through her symptoms and differential diagnoses, trying to determine whether it was one of:

- addiction
- bipolar
- depression
- schizophrenia
- anxiety.

Eventually they came up with the diagnosis of schizophreniform disorder. Her spark was dulled by the medications. Our family was in shock, and we were left with more questions than answers. Francesca had incredibly mild symptoms, like hearing voices at first, but the treatment became more traumatic than the problem being solved.

Chapter 7: ECHOES FROM OUR PAST

What could our family have done to prevent Francesca from going into hospital and having this diagnosis at 21? At this time, I had no idea why this happened to her and not the other three of us.

Why I Became a Neuroscientist

Sitting in that dimly lit room, listening to a prognosis that painted Francesca's condition as a lifelong struggle with little hope for recovery, I had a moment of realisation: *we clearly don't understand the brain because the treatment being proposed seems inappropriate for what her symptoms actually are.* It was at that moment that I mentally shifted gears—from being a pharmacist to someone determined to delve into the intricacies of the brain, to figure out what was really happening and how we could do things differently.

This led me to pursue a PhD in neuropharmacology, followed by a postdoctoral fellowship in neuroscience, and eventually running my own lab in San Francisco and now in Brisbane. For three decades, I committed myself to researching brain science, focusing on neural pathways and the development of more effective medications. But despite all this, something always felt incomplete, as if a crucial piece of the puzzle was missing.

Unbeknownst to me, a whole different field of research was unfolding in parallel, one that would eventually fill in the gaps and revolutionise my understanding of brain health.

Back in 1989, two researchers, Doctors Anda and Felitti, had embarked on a journey to explore the factors contributing to the risk of physical and mental health disorders. Their groundbreaking research wasn't published until nearly a decade later, in 1998. Known today as the Adverse Childhood Experiences (ACE) Study, their work delves into childhood trauma and maltreatment. This pioneering research has since been replicated globally and has significantly expanded our understanding of health and well-being.

For the past 8000 or so years of recorded history, there has been little progress in the treatment of mental health disorders, in stark contrast to the rapid development of new vaccines like those for COVID-19. In 1989, Francesca's experience was no different from that of others facing mental health challenges. Hospitals, armed only with the available treatments of the day, primarily used medications. Her regimen became a prolonged series of antipsychotic drugs that left her overly sedated and rigid when administered in high doses.

The antipsychotics Francesca was prescribed have origins back in the 1940s, and function by blocking dopamine receptors in the brain. Subsequent medications were merely slight chemical variations of these original compounds. Initially, the drugs alleviated her symptoms enough for her to leave the hospital, travel, complete a bachelor's degree, and lead an independent life. But while these medications made her feel "normal", they were gradually erasing the person Francesca used to be.

Adverse Childhood Experiences (ACEs)

> ***Trigger Warning and Cautionary Note:*** *The upcoming content dives deep into scientific findings that might be confronting, distressing or triggering, especially for those unfamiliar with the subject. While it's essential to understand these issues, navigating them can be challenging. If you find any of this overwhelming, please consider seeking professional guidance or discussing the content with a knowledgeable professional. Knowledge is powerful, but your well-being is paramount. If you feel more comfortable, consider undertaking this reflection with a trained professional or skip the rest of this chapter.*

Science has revolutionised our understanding of how genetics and early life experiences can significantly influence the likelihood of developing mental health disorders later in life. As our scientific knowledge evolves, we have a responsibility to incorporate these insights into the standard of care across all clinical settings and practices.

There were three pivotal moments that transformed my research journey to understand Francesca's struggles with mental illness. The first came when a colleague in the lab next door demonstrated that the adult brain is actually "plastic", or changeable. She showed me live experiments with an adult rodent exercising on a treadmill. What was amazing was that she could show how the neural connections, or synapses, in that brain were changing as they exercised. This went against what textbooks had been saying—that our brains stop changing after we turn 25.

The second eye-opening moment occurred in my own office while reading the work of Dr Bruce McEwen. He's well known for his studies on how stress hormones can change the way our brains are wired. This got me thinking about the intricate ways our experiences and stress levels can influence our mental health, opening new directions for my research. In everyday language, McEwen told us that our life experiences have a big impact on our brain. Not only does our brain feel stress, but it also plays a major role in how we deal with stressful events. Being resilient—able to bounce back after hard times—can affect whether we develop mental health issues like anxiety, depression, or addiction. Before McEwen's work, people thought the adult brain couldn't change. But he and some other scientists showed that our brains can change throughout our lives, influenced by hormones and other factors. This was revolutionary for both brain science and medicine.

The third moment was a game-changer, forever altering my trajectory as a neuroscientist. It was discovering the pioneering studies of Doctors Felitti and Anda, who were trailblazers in establishing the ties between adverse childhood events and subsequent health issues. This discovery was like finding a missing jigsaw piece I had been seeking for decades. They developed a tool known as the Adverse

Childhood Experiences (ACEs) questionnaire, first published in 1998, which consists of ten questions designed to assess childhood trauma.

The questionnaire breaks down the types of trauma into categories. The first five questions focus on personal experiences: physical abuse, verbal abuse, sexual abuse, physical neglect, and emotional neglect. The remaining five questions examine familial factors: having an alcoholic parent, a parent who has been a victim of domestic violence, a family member in jail, a family member with a diagnosed mental illness, or the absence of a parent due to divorce, death, or abandonment.

This tool enables healthcare providers to better understand the number and types of adverse experiences someone may have encountered before reaching adulthood, thereby offering crucial insights into their overall well-being. The findings from the ACEs study, corroborated by scientists around the globe, show that a higher ACE score significantly increases your risk for a variety of health issues, including not just mental health conditions like depression, but also physical ailments like heart disease and diabetes. What's startling is that a higher ACE score exponentially increases your risk for severe issues like alcoholism, obesity, bipolar disorder, and schizophrenia. It's crucial to dispel the myth that these issues are a result of personal failures or moral shortcomings.

Imagine showing up at a hospital for the first time and being told you've already had five heart attacks—that's what it feels like when someone is diagnosed with a mental health disorder after years of unaddressed traumatic experiences. It can seem overwhelming and hopeless, especially when told that you may have to manage these conditions for life.

But here's the thing—you're not alone in this struggle. One in four people will need mental health care at some point in their lives, largely due to adverse experiences in their formative years. Understanding these early life traumas and their long-term impact on our mental health is more vital than ever. It's about more than just treating the symptoms; it's about addressing the root causes to improve future outcomes.

Although 70% of people experience at least one traumatic event in their lives, most of us are unaware of the concept of ACEs. And while the DSM-5 (The *Diagnostic and Statistical Manual of Mental Disorders*, fifth edition, our current system for diagnosing mental health issues) lists as many as 300 different mental disorders, providing a comprehensive guide for healthcare professionals, it often doesn't delve deep enough into the origins of these conditions. This is why integrating an understanding of ACEs into both public awareness and clinical practice is essential for a more holistic approach to health and well-being.

Understanding ACEs often starts by examining our society, community, and our own family histories and life experiences. Dr Anda, one of the researchers behind the ACEs study, said that learning about ACEs changed how he interacted with his own teenage children. He found himself questioning why he spoke to them the way he did and whether he was unintentionally stressing them out. He realised that many of his behaviours were learned—either from society's expectations of parent–child relationships or from his own upbringing. This led him to understand that old habits need to be unlearned and replaced with new, healthier ones.

For me, coming across the ACEs study was a revelation. It helped me make sense of why Francesca developed schizophrenia. No wonder I always felt like there was something missing in my research, trying to uncover the root causes of mental health disorders by using only a neuroscience and medications lens. I had now found what I had been looking for over the previous 20 years. It's not just about diagnosing and treating symptoms; it's about understanding the root causes and how they can shape a person's mental health. This knowledge is essential for making meaningful changes and improving outcomes for those struggling with mental health issues.

💬 CASE STUDY – Lily

Meet Lily, a young woman in her thirties who grew up in a household plagued by domestic violence and substance abuse. From a young age, Lily was exposed to physical and emotional abuse from her alcoholic father, who would often come home in a rage and take out his frustrations on her mother. This constant state of fear and unpredictability had a profound impact on Lily's development and well-being. Despite the challenges she faced at home, Lily was a bright student and tried her best to excel in school. But the trauma she experienced at home often manifested in the form of anxiety and depression, making it difficult for her to focus and concentrate in class.

As she entered adulthood, Lily found that her childhood experiences continued to haunt her. She struggled with trust issues and had a hard time forming meaningful relationships. Her mental health issues worsened, and she turned to drugs and alcohol as a form of self-medication. When she hit rock bottom, Lily eventually sought help and entered therapy. With the support of her therapist and a strong network of friends and family, Lily was able to work through her trauma and learn to manage her anxiety and depression. She realised the importance of self-care and began to make positive changes in her life, such as eating healthier and engaging in physical activity.

Lily's story is just one example of the far-reaching effects that ACEs can have on an individual's health and well-being. However, it's also a testament to the resilience of the human spirit and the power of seeking help and support to overcome even the most challenging of circumstances.

Many people may not realise that a history of ACEs across generations can amplify our stress reactions. Every individual is unique in how they react to stress, largely because everyone has a different set of ACEs, resilience factors, and protective mechanisms. Think of it like a mathematical formula: your ACEs plus your resilience and protective factors equal your unique stress response. This is why some coping strategies or treatments work better for some people than for others.

But the traumas or ACEs we encounter in our early years are not our burdens to bear alone. Though they are part of our story, we have the power to decide how we proceed by gaining a deeper understanding of ourselves. Choosing courage over pain can not only transform our lives but also positively influence the lives of our children. Understanding this "formula" for stress can help us tailor more effective, personalised coping strategies. It can also guide parents in nurturing resilience and protective factors in their children, setting them on a path to handle life's challenges in a healthier way. This cycle of understanding and adaptation can have a profound, multi-generational impact, empowering us and our children to live more fulfilling lives.

How do ACEs Affect Parents?

Why do adverse experiences in childhood make it more likely for someone to have mental health issues later in life? It's because these experiences can change the way a young brain grows and works. Childhood is a formative period in which the foundation for mental and physical health in later life is laid. Negative experiences during these early years, like maltreatment, can have a lasting impact, shaping the "architecture" of the nervous system, as previously discussed. This in turn can lead to a myriad of health issues, from mental conditions like anxiety and depression to physical ailments and even a shorter lifespan.

In fact, on the extreme side of things, vast research shows how not being seen early on can lead to addiction, poverty, and crime. Youths don't end up offending or being involved in criminal activities overnight. Most often, they come from a home with absent parents or from an abusive household. Ninety percent of these young offenders have also been disengaged from education, services, and family from a young age. They then end up out on the street, hanging with the wrong crowd, and committing crimes. All these young people have the same thing in common: no one is looking after them. They haven't been *seen*. They don't have close relationships with a carer or parent. While developing programs for these children in the last few years, we have made incredible strides forward in understanding the individual's history and where they come from. As we now know, if children one to three years of age don't have the right amount of attention from their carers or parents, adverse changes in the brain are more likely to occur. Scans of these children's brains have shown decreased oxygen in very specific areas of the brain, namely the prefrontal cortex. This indicates decreased physical connections and demonstrates how drastically the brain architecture can be changed early on, essentially setting you up for the rest of your life. And these changes also happen across generations. The genetic lottery plays a part.

In contrast, children who experience healthy and close relationships with their parents and/or caregivers (even if they have a high genetic risk) in a healthy community and society that supports them gain more protection from mental health concerns and tend to perform better than those who don't. By receiving the attention they need, they build healthy neural networks. At the same time, a support system later can also make a big difference in overcoming the effects of ACEs—indicating that it's never too late. Healthy societies and strong parents in the home and proper attention and love may counterbalance these intergenerational effects.

During the first three years of life, a baby's brain is hardwired to grab the attention of carers or parents. This helps them acquire the proper skills (such as social, language, motor, and more). Beyond the first years of life, between the ages of 12 and 17, the pruning of many neural pathways takes place. And this means that the parent–child relationship is also very important throughout the adolescent years.

It's important to note that the relationship between ACEs and brain development is complex and multifactorial. Factors such as genetics, environmental stressors, and access to support and resources can all play a role in determining the extent to which childhood experiences impact brain development and health outcomes.

To be clear, it's not *just* about these experiences. Things like your genes, the stress you're under, and the support you get from family and friends, community and society also matter. But when a child goes through tough times, it can mess with how they deal with stress for the rest of their lives. Their bodies might start reacting as if they're always in "emergency mode", which isn't good for the brain. These ongoing stress reactions can mess with normal brain growth and can make it more likely for someone to have health problems when they get older. It's not just mental health issues like depression, anxiety, or addiction; it can also lead to physical problems like heart disease or diabetes.

Some studies even show that these experiences can change how our genes work, affecting the brain's wiring and making it more likely to have chronic health problems. Researchers have found that people who had a lot of bad experiences as kids are more likely to suffer from issues like depression, anxiety, post-traumatic stress disorder (PTSD), and addiction as adults. Understanding that ACEs can have a long-lasting impact on our brains and health is important. It helps us see that mental health issues aren't a sign of weakness or bad character; often they're connected to experiences people had no control over when they were young.

Imagine building a house. First, you lay the foundation, right? In the same way, our genes set the groundwork for our nervous system. These genes have been passed down through generations and they create proteins that serve as the building blocks for our nervous system. This system, in turn, shapes how we understand ourselves, especially when influenced by our experiences and environment. The thing is, our nervous system's "architecture" is always changing based on what we go through in life. And the tougher those experiences are, especially when we're young, the more they affect us. Many times, these experiences are buried deep in our memories, kind of like how we might put new wallpaper over old wallpaper in a house instead of stripping the old away and starting fresh. To make it even more complicated, some of the things we react to aren't even from our own life experiences.

ACEs Affect Overall Health

ACEs can have lasting impacts on health and well-being across the lifespan. It's important to note that ACEs are not a direct cause of these negative outcomes, but rather they increase the risk of developing them. We do not understand the precise brain mechanisms that generate the 300 differential and overlapping diagnoses in the DSM-5 yet! Nevertheless, it's clear that ACEs can have a significant impact on an individual's health and well-being throughout their life. Research has shown that exposure to ACEs can increase the risk of developing a variety of physical and mental health problems, including:

- Cardiovascular disease. Chronic activation of the body's stress response system as a result of ACEs can increase the risk of developing heart disease and stroke.
- Chronic pain. Childhood trauma has been linked to an increased risk of chronic pain conditions, such as headaches and back pain.
- Mental health problems. ACEs can increase the risk of developing depression, anxiety, PTSD, substance abuse and addiction.

- Obesity and poor nutrition. Childhood trauma has been associated with poor dietary habits and an increased risk of obesity.
- Immunity and infections. Childhood trauma can affect the functioning of the immune system, making it more difficult for the body to fight off infections.
- Chronic physical health problems, including, diabetes, and certain types of cancer.
- Impairments in cognitive function and academic achievement.
- Increased risk for crime, imprisonment, violence and victimisation.
- Poor health behaviours such as smoking, lack of exercise, and unhealthy eating habits.
- Interpersonal difficulties and relationship problems.

Why are the ACEs results not being included in programs to prevent and improve the treatment of mental illness? Why are they not being shouted from the rooftops like other breakthroughs, and not becoming a mainstay in medical and healthcare practice? Why do they remain in silence? However, there's a silver lining to this cloud. As history has shown us, with increasing awareness and relentless advocacy, whispers can turn into empowering roars. We stand at a pivotal moment, where understanding and acceptance of mental and brain health as a crucial aspect of overall well-being is gaining momentum. As more individuals, professionals, and institutions become aware, there's hope that this will herald a new era in which mental health is universally recognised and addressed, not as a luxury or an afterthought but as a **BASIC HUMAN RIGHT**.

Effective parenting is anchored in mental well-being, and understanding the scars of childhood experiences is pivotal. With mental health standing as the cornerstone of nurturing parenting, it's essential to address the dark clouds of ACEs that many bear. Despite the discomfort and societal stigma surrounding discussions of childhood maltreatment, shedding light on it is paramount, not just for personal healing, but for broader public health and the wellness of subsequent generations.

Silencing or sidelining these discussions only reinforces the cycle of pain and hampers potential interventions. The urgency of this dialogue is underscored by staggering statistics: estimates suggest that between 20% and 70% of adults have faced at least one ACE, with up to 20% enduring four or more. The reverberations of ACEs extend deeply into adulthood, affecting mental and physical health, relationships, and even parenting styles. For the sake of our children and their futures, it's imperative we confront these shadows, fostering a society that prioritises mental health and breaks the chains of inherited trauma.

By choosing to engage with this book, you are taking a proactive step towards breaking the cycles. You're not only bringing about change for yourself but also paving the way for future generations to live without the weight of past adversities. Through understanding and action, we can ensure that brain health becomes everyone's business and not someone else's problem.

Time for Reflection

If you feel more comfortable, consider undertaking this reflection with a trained professional. Think about your childhood and any experiences, whether positive or negative, that left a lasting impression on you. How have these experiences shaped your approach to parenting or relationships?

Chapter 8

THE TECHNOLOGY FACTORS

The Serious Challenges for Parents in the Digital Age

In Brief

It's undeniable that technology has transformed our daily lives. At the onset of the social media age, people were finding organ donors online. They were reconnecting with long-lost loved ones. We were quickly deceived into thinking social media was good, even to kill time as we waited for the bus or to be called in for an appointment. But as engineers at technology companies began competing for our attention, they created a monster—in more ways than one. The digital revolution, with its smartphones, social media, and quick video deliveries, has left a serious impact on our children's physical and mental well-being (and our own!). There are even babies of only 18 months of age able to navigate YouTube now. While in some ways impressive, it's not all good. Everywhere you go, children are sitting in front of tablets and screens, whether in a doctor's reception area or a restaurant, or at home. This problem is now at pandemic levels and is starting to outweigh all other parenting priorities. ***It is happening here and to ALL children with access to devices and the internet!***

As we grapple with the incessant demands of contemporary living—from work pressures and financial stresses to social obligations and the challenges of parenting—many of us instinctively turn to our smartphones as a refuge. The allure of scrolling through our digital screens promises a brief escape from the world around us. Yet, paradoxically, this very act, which we believe provides relief, often exacerbates our stress. Amidst the continuous hum of notifications, social media updates, and the vast digital world, we find ourselves even more entangled in the web of stress, a barrier to being an effective parent.

In a world where even technology industry leaders limit their own children's screen time, it's absurd that we are handing over these devices to kids who can't even vote or drink legally. Why are we allowing our children to serve as test subjects in a digital experiment with unknown long-term consequences? It's high time we take collective action to put our children's safety and well-being above all else. They have an opportunity to be a child only once; why would we want them to leave behind the joys of childhood too soon? The only real answer is to see the new technology and devices as a weapon or a cigarette. Would we let our children use *them*? For most, it's a resounding no.

This is an emergency for the safety of our children. It is going to take more than a traditional village to solve. It's infuriating that parents must shoulder this burden largely alone, thanks to a lack of protective laws against the technology giants and people exploiting the opportunity to profit from our kids' screen time. And let's face it—parents are already stretched thin. The pressure to keep up with the latest technology trends, under the belief that it's essential for their children's future success, only adds to the stress.

Now, none of this is to make you feel guilty. As a parent myself, I understand how we easily fall into the trap of blaming ourselves. Rather, this information is to help you gain awareness and create limits around social media and technology so your child can truly thrive. You may also want to approach these boundaries or limits as a family, since reducing screen time has immense impacts on mental health for all ages, and by having boundaries together, you can foster in-person bonds as opposed to one-way relationships on social media.

The Impact of Technology on Brain and Child Development

How do smart phones and social media change children's minds through the remarkable ability of the brain to reorganise its neural connections in response to experiences, learning, and environmental stimuli? Let's dive in.

Affects Mental Health. As you might have gathered by now, social media is a possible contributor to the rise in mental health issues in our society. Excessive smartphone use may increase various mental health issues in children and adolescents, the research evidence is starting to show. The constant comparison with others and seeking of validation can lead to a range of problems like anxiety, depression, eating disorders, addiction, self-harm, and sleep disturbances. As parents, we must be vigilant.

Leads to Addiction. Many of us are on social media apps for well over two hours a day. The effect of these apps on the reward centre of our brain makes them as addictive as drugs and alcohol. This is an *emergency*, and it's happening right now. We, as parents, need to be aware of how damaging the effects are on the developing brain. Many of us absentmindedly use social media. We scroll without thinking about or knowing what's going on behind the scenes. I am guilty of this. A recent investigation by *Guardian Australia* newspaper found more young adults are facing depression, anxiety, debt, and relationship troubles due to early gambling habits. Online stats show that in 2022–23, there was a 16% rise in young people of 24 and under asking for help with gambling problems. In Victoria alone, around 600 of those help calls came from young people aged 15 to 24.

The algorithm that determines what we see on these apps was created by humans but runs on its own agenda according to what we feed it. It shows us more of what it *thinks* we want to see, fuelling a behind-the-scenes money-making machine, and in turn, it becomes even more addicting. Even the design of each app is thoughtfully created to keep you engaged and interested. It'll show you fake news that it thinks you like, whether that news is real or not. It can trick even the smartest individuals among us into believing we aren't good enough or pretty enough because we're comparing ourselves to what we're shown on social media. I highly recommend that you view the documentary *The Social Dilemma* on a streaming platform, as it highlights how technologists learned how to hack our brain to make us addicted to our devices. It is shocking; I even find myself addicted to devices and technology, as they are required for all parts of work and life. It often feels like the issue of too much technology and screen time is too big for us to tackle.

Social media activates the reward centre of the brain. For pre-adolescents, this can be particularly addictive. At this age, we're supposed to be looking to our peers and learning how to gain social rewards in person. Instead, we're learning these cues online, a very different reality compared to face-to-face interactions. We have a constant stream of stimulation through notifications, apps, and social media updates. This flood of information can trigger changes in the brain's neural pathways, shaping the way we process and retain information, and making it difficult for us to come off them. Children's brains do not have the capacity to handle the deluge of information delivered by smartphones, social media, and devices.

Steals Our Attention. Technology shortens our attention spans. Constantly switching between tasks on our mobile phones, including scrolling, is exhausting and depletes the brain's valuable resources. For the developing brain, this can make learning difficult, as we lack the concentration to take in information. It can also make relationships difficult, as we struggle to pay attention to in-person interactions.

Stunts Our Learning. In the past ten years, the Australian education system has seen a large drop in outcomes. But recently, smartphones were banned in Australia due to this huge decline and what we know about social media's impact on mental health and brain health. The first day after the ban was put in place, a teacher mentioned in the news that students seemed more focused and were actually paying better attention to their lessons in class.

Reduces Our Self-Esteem. In the time before social media, we had only our peers or immediate world to compare ourselves to. Now we have billions of people to compare ourselves to every day. When we see people online who appear flawless, always happy, and leading perfect lives, it's natural to feel like we're not measuring up. And this can not only stunt us socially but also create a social media addiction, where we spend more time in the virtual world than in the real world.

Supports a Sedentary Lifestyle. Social media usage supports sedentary behaviour. We spend more time sitting than ever before, contributing to obesity, heart disease, diabetes, and more. Excessive video gaming can also lead to a hyperaroused state, leading to further difficulties in concentrating, paying attention, managing emotions, and controlling impulses.

Disturbs Our Sleep. Blue light, which is emitted from our smartphones and digital devices, has the power to suppress melatonin. As a result, it can significantly impact sleep. For our children, this can have extreme implications for brain development, impacting their learning, memory, and creative-thinking capabilities.

Facilitates Grooming and Online Danger. The digital world also presents risks related to online grooming and child safety. As parents, we must be proactive in educating ourselves and our children about online safety measures. Child abuse and sex trafficking have become increasingly urgent concerns in today's world. We'll delve deeply into these topics in the upcoming chapter. Not all is what it seems online. Your child's new "friend" might not be their age, despite how they may present. In fact, they could be a 60-year-old man sitting only a few blocks away, trying to lure innocent children into his web (I'm not kidding). Child abuse and sex trafficking are taking place behind the scenes online. Individuals will try to coax children to meet them outside of the online space, where the worst nightmare for any parent can happen.

Enables Cyberbullying. Many of us have probably heard stories about cyberbullying happening unbeknownst to the parents, since it takes place online. Because of this, it can fly under the radar and evade authority figures, such as teachers or other parents. And unfortunately, this is a large factor contributing to anxiety, depression, and high distress. In extreme cases, young children have been driven to suicide by cyberbullying.

Technology Makes a Poor Babysitter

I've seen countless parents do it, and I'm as guilty of this as well; when grocery shopping or waiting for an appointment, the easiest way to keep our kids quiet and behaved is to throw a tablet or phone in their lap. But I'd advise against this. While tempting, this has serious effects on your child's brain health and well-being. In fact, right now, most experts are advising parents not to allow their children to use phones until they are at least 13 years of age.

After all, technology companies aren't thinking of the health of your child; they're in it to make money. If you've ever watched the documentary *The Social Dilemma*, you'll also know that many technology executives and CEOs don't allow their children to use phones. They know it's not good for them.

Identifying Addiction to Devices and Signs of Withdrawal

A word of warning: initially, distancing yourself from technology is likely to be a challenging endeavour, akin to quitting smoking, drinking, or gambling. As we are deeply intertwined with our devices, sudden disengagement can trigger a range of emotional and behavioural symptoms like those experienced in other forms of addiction. Recognising these symptoms of withdrawal from devices is essential for successfully implementing a new routine for technology and screen time within your household. Like overcoming other forms of addiction, taking that first step is usually the most challenging part. Armed with the understanding of what to anticipate, you'll be better equipped to stay the course, even when the going gets tough.

Here are some signs of addiction to devices you might observe:

> **Withdrawal from Social Situations.** Isolation from family and friends in favour of spending time online or engaged with technology.

Neglect of Responsibilities. Ignoring work, school, or household duties due to the compulsion to be online.

Lack of Self-Control. Despite knowing the adverse effects, an inability to cut down on technology use.

Loss of Interest. Reduced participation in previously enjoyed activities that don't involve technology.

Anxiety. Feelings of unease or panic when separated from devices or unable to check social media or messages.

Mood Swings. Rapid shifts in emotional state, ranging from euphoria when online to sadness or emptiness when offline.

Here are some signs of withdrawal following addiction to devices you might observe:

Irritability. A heightened sense of frustration or agitation or physical violence, throwing things, swearing and screaming when unable to access technology.

Anxiety. Feelings of unease or panic when separated from devices or unable to check social media or messages.

Mood Swings. Rapid shifts in emotional state, ranging from euphoria when online to sadness or emptiness when offline.

Depression. A sense of hopelessness or despair may set in, particularly if the individual perceives their "real life" as less rewarding than their online interactions.

Online Child Exploitation Is Happening Worldwide

As you know by now, the brain is shaped by the environment and experiences, especially in our youth. These impressionable times can be significantly impacted by the use of tablets, smartphones, and

Chapter 8: THE TECHNOLOGY FACTORS

other devices. Ultimately, we do not want our children to learn from screens, which can leave them unable to learn to read human faces in the real world (having serious social, emotional, and mental impacts!). And yet, all of this is happening as this chapter is being written. Now, if the above information doesn't ring many alarm bells for you, then what about online child exploitation, sextortion, pornography being viewed by children as young as six and seven years old, grooming, and other exploitative acts happening online? The truth is that the safety and security of our children are what's at stake, especially as they navigate particularly vulnerable ages.

Addressing the issue of online child exploitation is unquestionably a difficult and sensitive topic, yet its urgency has never been greater. We are at a unique intersection in history, where a confluence of factors has intensified the risks for our children. The ubiquitous presence of smartphones, social media platforms, and online games provides exploiters with unprecedented access to potential victims. Simultaneously, parents, already stretched thin by the demands of work and life, are often distracted by their own devices, inadvertently increasing the vulnerability of their children. In this complex landscape, unscrupulous individuals find ample opportunities to prey upon the young, turning the digital age into a profitable enterprise for illicit activities. The importance of tackling this issue head-on cannot be overstated, as the stakes have never been higher.

An expert in online child exploitation once told me that the odds are higher that your child has encountered inappropriate content online than not. The most effective approach, then, is to operate under the assumption that your child may be exposed to such content. This proactive stance allows you to put preventative measures in place beforehand, rather than resorting to punitive actions after discovering that they've seen or done something they shouldn't have.

Grooming: Learn the Signs

Grooming is a calculated process where an individual employs a manipulative strategy to gain the trust of a child or young person, ultimately leading them into sexual abuse and exploitation both online and offline. Groomers invest considerable effort in building relationships, strategically garnering trust and rapport by understanding the likes, dislikes, and vulnerabilities of their targeted victim. Armed with this intimate knowledge, they work to break down the child's natural defences, establishing a bond that affords them control and continued trust. Sexual content and physical contact are gradually introduced into the relationship.

To carry out their manipulative tactics, groomers often exploit digital platforms like mobile phones, social media, and the internet. They may ask the child to keep their relationship a secret, making the crime more difficult to detect. Given the anonymity and reach of digital platforms, understanding and combating grooming has become a crucial parenting skill in the 2020s. Child exploitation isn't limited to face-to-face situations now; it can also occur in the digital world. This means our kids might be vulnerable even if they don't physically encounter a perpetrator. Exploitation happens when someone preys on a child's innocence, making them do things they might not be fully aware of, often for the exploiter's gain or prestige. And sometimes it might look as if the child was willing, even if they weren't.

What would be the ultimate safety net? A ban on smartphones and digital devices until they are completely safe for children's developing brains—and the onus would be on the developers, not the parents. An analogy: when we develop medications for blood pressure or diabetes, the drugs have to go through a rigorous process to say they are safe for humans. This is often a 10- to 20-year process and costs about a billion dollars. To be marketed to the Australian public they must be approved by the Therapeutic Goods Administration (TGA) or, in the US, by the Food and Drug Administration (FDA).

Chapter 8: THE TECHNOLOGY FACTORS

This is not a foolproof process, and even these medications can be marketed to children without being tested. But it's better than the process with technology, whose developers have been able to market to a global population without *any* regulations. While we prefer to turn away from these facts and think bad things are not going to happen to *our* children, I am letting you know that such exploitation is happening every second of every day. With a lack of regulations within the online world, as parents, we are the only true safeguard. It's up to us to put down our phones and avoid relying on technology for babysitting.

Let's face it—there's no sugarcoating this. The situation demands action, and it starts with us, the parents. If we don't grasp the ins and outs of technology, if we can't navigate privacy and security settings, and if we are unaware of what our kids are consuming online, then we must reconsider granting them access to smartphones and social media altogether. This isn't about taking away their fun; it's about protecting their futures and well-being.

Parenting challenges have escalated to include not only traditional worries but also a myriad online threats, one of the most insidious being the risk of grooming. Understanding how to protect children from grooming is increasingly becoming a vital parenting skill for the 2020s. Parents face an unprecedented set of challenges and need to tackle sensitive topics like sex education and online safety at younger and younger ages.

The United Nations defines child sexual exploitation as a situation where a child is abused, and in return, the wrongdoers gain something—it could be money, social status, or even political benefits. They emphasise that this exploitation is not just forceful but harms the child's health, growth, and learning. The issue of child sexual exploitation (CSE) is often misunderstood as a problem of limited choices—be they social, emotional, or economic. Initially, victims may believe they are in control and making their own decisions, only to find out too late that they are being manipulated.

Contrary to popular belief, Australia is not immune to this issue. Many assume that CSE happens to other people's children or in other countries, but the reality is much more unsettling. CSE can affect any young person, including those who are in foster care, have learning or physical disabilities, are homeless, or have other vulnerabilities such as mental health issues or substance abuse.

The scope of this problem is not confined to developing nations. Over 10 million children and young people are being sexually exploited worldwide, and comprehensive research shows that CSE is a global issue affecting children irrespective of their socio-economic background, age, or gender.

The UK government estimated the economic impact of CSE on its economy to be around AUD5.1 billion, highlighting the far-reaching ramifications of this issue. Understanding and addressing the risk of sexual exploitation has thus become an urgent priority for societies around the globe.

In Australia, there's a notable lack of recent and thorough data on CSE. The last major study sponsored by the federal government was in 1998 (Grant, David & Grabosky, 2001, "The Commercial Sexual Exploitation of Children"). The Royal Commission into Institutional Response to Child Sexual Abuse, completed in 2016, also falls short in this regard. It offers only two recommendations specifically pertaining to CSE, illustrating a systemic lack of focus on this critical issue.

Traditional "top-down" approaches to parenting, where information flows one way from parent to child, are becoming increasingly inadequate for preparing children to think critically, make informed decisions, and protect themselves in an ever-complex world. As discussed previously, becoming educated, opening our eyes wide and developing parenting skills fit for the digital age will assist with identifying the early signs of CSE and online grooming.

Chapter 8: THE TECHNOLOGY FACTORS

Types of Child Sexual Exploitation

Inappropriate Relationships

In this form of exploitation, the abuser usually holds disproportionate power—be it physical, emotional, or financial—over the young person. The youth often believe they are in a genuine or loving relationship with the abuser. This exploitation often targets individuals who lack a positive, nurturing adult presence in their lives.

Organised Exploitation and Trafficking: Commodification of Youth

Here, victims are moved through criminal networks, often between different towns and cities. The young person is typically forced or coerced into sexual activities with multiple individuals. This form of exploitation treats children as commodities, sometimes even branding them with tattoos to indicate "ownership". In Australia, case studies have shown young victims being "tagged" in this manner.

Peer-on-Peer Exploitation: Youth Targeting Youth

In this scenario, the abuser is roughly the same age as the victim, and everyone involved is usually under 18. Common settings for this form of exploitation include schools, youth clubs, and sports clubs. It can often occur within the context of "gang" activities.

Exploitative Relationships: The "Boyfriend/Girlfriend" Model

In this form, the abuser grooms the victim by initiating what appears to be a normal relationship. They may shower the young person with gifts and affection and often meet in public places like shopping centres, fast-food outlets, and hotels. As the relationship progresses, it takes a dark turn, becoming physically, sexually, and emotionally abusive. The victim may be forced to attend parties and engage in sexual activities with multiple individuals. The abuser may also threaten the victim with violence if they attempt to seek help and may coerce them into recruiting other potential victims.

Time for Reflection

Over the next week, track your daily screen time. You can use built-in features on most smartphones or download a specialised app to help you. At the end of the week, calculate your average daily screen time. Does the number surprise you?

After each prolonged screen session, take a moment to reflect on how you feel. Are you more anxious, stressed, relaxed, or neutral? Do certain activities (like social media browsing or reading news) make you feel different than others (like watching a movie or video chatting with family)?

Reflect on the times when you were engrossed in your screen. Were there moments when your screen time took precedence over engaging with your children? How did those moments make you feel afterward? Be honest with yourself. Are there times when you feel restless without checking your phone? Do you find it challenging to stay away from screens during family time, meals, or before sleeping?

Chapter 9

LOST CONNECTIONS

From Yourself, Your Child and Others

In Brief

> ***Please note***: *The first paragraph of this chapter describes a story of recovery from self-harm and a suicide attempt. Please consider whether you would like to read this. If not, please skip the first paragraph. If you require support, a helpline is available from Lifeline (13 11 14), and support can be provided by counsellors and other mental health professionals.*

One day, a young man called Joey woke up in a hospital's ICU. His mum told him there'd been a terrible accident. In fact, he had tried to take his life, and as a result he had lost a leg. Joey's tough experience and his remarkable recovery is shared in the documentary film *The Great Separation: Ambition for a Better Life Together*. In the film, Joey talks about what led to his decision and what he learned that made him feel better. After breaking up with his girlfriend, Joey had felt alone, pushing away his mum and friends. Everything came to a head on Christmas Eve in 2019. Joey's journey of self-discovery ultimately led him back to the community he thought he had lost forever. I had the pleasure of being part of the film that explores the impact of loneliness, disconnection, impact of social media, and lost community on our lives and the steps we can all take to reconnect.

Social isolation and loneliness aren't just about feeling down; they can genuinely hurt our mental and physical health and our ability to be effective parents. In Australia, these feelings have become major health concerns, and the COVID-19 lockdowns might have made things worse, especially for those living alone. Here's the thing: social isolation and loneliness aren't the same. Social isolation means you don't have many social contacts, while loneliness means feeling disconnected, even if you're around people. They can happen together, but not always. Raising a child isn't only a parent's job. It often involves grandparents, other parents, and the wider community. As the old saying goes, "It takes a village to raise a child." Now, imagine the uphill battle faced by parents who don't have the support of that "village". Loneliness or social isolation can be massive barriers to effective parenting.

People are busy, stressed and immersed in their digital devices. Instead of chatting in the elevator today, we take out our phones, perhaps to avoid any awkward small talk or to quickly escape reality, even if that's for a mere 10 or 20 seconds. Instead of reading a book, many people scroll for hours on end. Social media tugs at our attention and can even be a full-blown addiction. Social media also flashes bright colours and is designed to vie for our attention, and more and more people are spending time in the digital world as opposed to face-to-face.

So, is this a problem? Well, we know connections are essential for optimal brain health. And when it comes to parenting or deep connections, physical affection is part of the mix. This means that face-to-face connections should still be prioritised whenever possible. The truth is we are wired to connect. It is essential for our survival. Our ancestors needed a sense of belonging and connection within their tribe to ensure they wouldn't face rejection, which often meant death. But today, the idea of social connection is becoming blurred.

We might have 1000 Facebook friends, yet are they really all our "friends"? Are we really connecting with all of them in a deep and profound way that fosters good brain health? Probably not.

Whereas only a few decades ago, everyone knew their neighbours, today we don't. Many of us don't know the family that lives two doors down. And perhaps this lack of face-to-face connection is one of the key factors driving this uptick in mental health disorders that we've seen in recent years.

Social Connection and Cooperation Are Fundamental

We are a social species, and social connection and cooperation are fundamental aspects of human behaviour influencing everything from our personal relationships to our societal institutions. In recent decades, Western societies have experienced an increasing trend of disconnection, driven by several key cultural, economic, and technological shifts. The advent of the digital age, while providing unprecedented opportunities for global connection, has ironically fostered isolation, as face-to-face interactions are often replaced by virtual ones. Social media platforms, though designed to bring people together, have often resulted in comparison culture and feelings of loneliness and inadequacy. Economic changes, including the rise of precarious and gig-based work, have also led to the erosion of community-based workplaces, further reducing opportunities for real-world social interaction. Meanwhile, urbanisation has separated us from nature and the communal lifestyles of our ancestors, and consumer culture has encouraged individualism and competition over community and cooperation. These trends have left many feeling disconnected from others, their environment, and even themselves, contributing to the growing mental health crisis observed in these societies.

When parents feel alone or unsupported, it becomes more challenging to meet the demands of raising a child. And here's something alarming: feeling isolated or cut off from others can have health impacts like chronic diseases. This isn't just an abstract concern; even before the pandemic hit, loneliness was already a significant issue in Australia. Recent data has shown that approximately one in three Australians have expressed feelings of loneliness. When we recognise the profound health risks and understand how these feelings can interfere with effective parenting, it becomes evident that building and nurturing our communities is more essential than ever, not just for the well-being of parents, but for the holistic development of our children.

Before we dig into more practical strategies in Part 3, let's take a moment to recognise that social connection plays a major role in reducing stress that we must not underestimate. When stress happens, many of us separate ourselves from our friends and family. We might withdraw or stay in bed. We disconnect and become anxious at the idea of going out to meet people. Then we feel lonely, sad, and maybe more stressed. In fact, many of us likely experienced this throughout the recent pandemic (and maybe still do). Lockdowns not only impacted us throughout those difficult years but may still be having repercussions on some individuals' well-being. Put simply, when stressed, we often end up putting a wall up between ourselves and our greatest support systems. Yet, from what we understand about human nature and the brain, we *need* meaningful connections. Without it, both our physical and mental health drastically decline. We need each other to navigate grief, stress, and other negative emotions. Our community helps support us through these tough times. Research even shows how positive social support makes us more resilient to stress, helping protect our brains from various functional and mental health consequences.

Social Connection Buffers Stress

As parents, it's easy to feel overwhelmed and under-supported. You're far from alone in feeling this way. Our societal and economic structures often propagate the myth of individualism—that we can, and should, make it on our own. Research studies and observations of flourishing communities suggest otherwise. In fact, they underscore the critical importance of supportive relationships as key protective factors for brain health and well-being. These networks of care not only enable us to thrive as individuals but also foster healthier, happier generations to come. In many communities, grandparents helping to raise kids adds a special touch of love, tradition, and history. It's not just about lending a hand; it's about sharing old family stories, teaching important life lessons, and keeping family traditions alive. Grandparents have a different way of caring, mixing experience with love, and sometimes a bit of spoiling. For kids, this means getting to know their family roots better and hearing fun stories about their parents as children. These special connections with grandparents create lasting memories and bring families closer together.

Many authors have written about the value of social connections. In Dan Buettner's book *The Blue Zones*, he explores several regions worldwide known for their extraordinarily long-living inhabitants, including the islands of Okinawa, Japan, and Ikaria, Greece. In these "Blue Zones", residents commonly live active, healthy lives into their hundreds. Buettner, through extensive research and interviews, uncovers the lifestyle, diet, and social elements contributing to the longevity of these populations. In Okinawa, for instance, a diet rich in vegetables and soy, along with a strong sense of community (or *moai*), seem key to their long lives. In Ikaria, a combination of a Mediterranean diet, regular physical activity, afternoon naps, and a close-knit community all contribute to a high rate of nonagenarians and centenarians. The book suggests that by adopting some of these practices, people elsewhere can also improve their health and potentially extend their lifespan.

Drawing from the powerful insights offered by Johann Hari's book *Lost Connections*, it becomes clear that our societal fabric and lifestyle significantly contribute to our mental health. Hari argues against the notion that depression and anxiety stem solely from chemical imbalances in the brain. Instead, he identifies nine real causes, primarily rooted in various forms of disconnection—from meaningful work, other people, values, and even our own future. The proposed solutions suggest that addressing these mental health challenges requires societal changes more than merely individual ones. Dr Mark Williams further extends this perspective in his book *Connected Species: How Understanding the Brain Will Save the World*. He details how our evolutionary history as social animals has shaped our brains, behaviour, and societal structures. Survival and reproduction in our early ancestors hinged upon living and working cooperatively in groups. Our biological evolution, especially the development of our prefrontal cortex, reflects this deeply ingrained social nature, enabling us to understand, empathise with, and predict the behaviour of others. Thus, forming social connections and fostering cooperation are central to our human identity and survival.

Loneliness and How It Affects the Brain

Feeling lonely doesn't always mean being alone. Research has shown that the feeling of loneliness and the degree of actual social isolation (that is, how often you meet or talk to people) aren't strongly connected. So, just increasing the number of social meetings might not cure this feeling. A significant portion of people experience loneliness. It's even more prominent among older adults, and those struggling financially feel it even more intensely. And here's why this matters: feeling chronically lonely isn't just bad for our mood; it's bad for our health and this affects our ability to be effective parents. Studies have linked persistent loneliness to heart issues, cognitive problems, depression, and even a shorter lifespan. In fact, the health risks of prolonged loneliness can be compared to the risks of long-term smoking.

With these concerning effects, researchers are diving deep to understand how our brains process loneliness. They're studying animals and humans to grasp how our minds and bodies respond when we feel isolated. They're hoping to find ways to help and support those who feel lonely, ensuring they lead healthier, happier lives.

Loneliness isn't just an emotional experience; it has tangible effects on brain function. For instance, lonely individuals often find it hard to trust others. Studies reveal that when such people participate in trust-related activities, specific brain areas linked to emotions (like the amygdala) are less active than the corresponding areas in those who aren't lonely. Another interesting find is in the brain's "mirroring" function, which causes us to subconsciously mimic others. Lonely individuals exhibit heightened activity in mirroring-related brain regions, yet they struggle to synchronise their actions with others'. Additionally, older adults who are lonely and show signs of depression display different brain reactions, particularly when asked to control impulses. Overall, loneliness seems to influence our brain's behaviour in social contexts and impulse control, serving as its way of responding to feelings of isolation.

Social Prescriptions: A Fresh Take on Parenting for Health and Well-Being

When we think of health, we often picture doctors, pills, or hospitals. But what if health could also be about volunteering, joining a local book club, taking an art class, or going for community walks? That's the idea behind "social prescriptions"—a refreshing way to look at our well-being. Unlike typical doctor's advice that might focus on medicines, social prescriptions are all about activities that connect us with others and make us feel good.

Imagine a doctor suggesting you join a gardening club, take a painting lesson, or participate in a local yoga class. These aren't just random activities; they're meant to boost our overall happiness, well-being, and sense of community. Such activities can be especially important in today's world, where many of us might feel alone or disconnected. By engaging in these group activities, we get to interact with others, learn something new, and break the monotony of our daily routines.

But why are these social prescriptions becoming more popular? Firstly, they look at health in a broader sense, understanding that our mental and social well-being is just as important as our physical health. They also directly tackle the feelings of loneliness many people experience today. Plus, being involved in these activities makes people feel empowered, taking an active role in their well-being. And the best part? Many of these activities are affordable or even free.

However, there are challenges too. For these prescriptions to work, doctors and community groups need to work together. This would ensure that when a doctor suggests an activity, it's available and easily accessible to the patient.

Time for Reflection

Remember, the goal of this exercise is not to judge or criticise, but to become more aware of our social patterns and feelings. This understanding will empower us to make positive changes and foster deeper, more meaningful connections.

1. Personal Connection Inventory

- List down the names of five people you are close to you.
- When was the last time you had a face-to-face interaction with each of them?
- How often do you engage in meaningful, non-digital conversations with these individuals?

2. Digital vs Physical

- Track your social media usage for a day. How many hours did you spend?
- How did you feel after each session? Connected, lonely, indifferent?
- Compare this with how you feel after an in-person catch-up with a friend. What are the significant differences in feelings and quality of connection?

3. Understanding Loneliness

- Think of a time when you felt surrounded by people but still felt lonely. Why do you think that was?
- Conversely, think of a time when you were alone but didn't feel lonely. What contributed to that contentment?

4. Village Reflection

- Reflect on your current "village". Who makes up your support system?
- Are there elders or grandparents or community figures you look up to for guidance or support? How often do you interact with them?

5. Work Environment and Community

- Do you feel a sense of community at your workplace or within your profession?
- How has the nature of your job impacted your social connections?

Chapter 10

LIFESTYLE FACTORS

The Unseen Impact of Poor Diet and Lack of Exercise on Parenting

In Brief

When it comes to parenting, there are two big pieces of the puzzle that many of us don't think about much. They are what we eat and how often we move our bodies. These aren't just about looking good or staying in shape—they directly affect how we feel from day to day. The overlooked reality is that poor food choices and a lack of exercise can silently erode a parent's ability to effectively care for their children.

Firstly, amidst the chaos of school runs, bedtime stories, and endless activities, it becomes all too easy to grab a quick fast-food meal or let that evening walk slide. Yet such habits chip away at our energy reserves. Consuming foods low in nutritional value can leave parents feeling sluggish, while the lack of physical activity can result in feeling constantly drained.

Ever notice how you feel after a good meal or after some exercise? Ever notice how it puts you in a better mood? That's your brain and body telling you something! The food we eat can either give us a pick-me-up or make us want to slump on the couch. Eating a lot of junk or sugary stuff might give a quick buzz, but it can also make us feel grumpy or tired later on—not the best when trying to keep up with lively kids.

Exercise, too, does more than just keep us fit. Whether it's a jog around the block, a few yoga moves in the living room, or just dancing around with the kids, moving helps lift our mood. It releases these feel-good chemicals called endorphins, which are like Mother Nature's happy pills. They help shake off stress and give us a brighter outlook on things.

But here's the tricky part: With everything parents have to do, like school runs, house chores, and bedtime routines, it's tough to eat well and get regular exercise. It's so easy to grab a snack on the go or say, "I'll work out tomorrow." But over time, this can make us feel worn out and more moody.

In short, food and exercise are secret weapons for parents. They don't just keep us healthy but also help us be at our best for those little ones who rely on us. So, in a later chapter we dive deeper into this topic. We'll explore some easy ways to make these two puzzle pieces fit better into our busy lives. After all, taking care of ourselves helps us take even better care of our kids!

Exercise and Parenting: The Wellness Link

In simple terms, our bodies are built to move and to be active, just like our ancestors were. But nowadays, many of us spend a lot of time sitting or immobile. Health experts like the Center for the Developing Child (CDC) tell us that we should be exercising about 150 minutes a week to stay healthy. However, many people aren't doing that. Not moving enough can not only make our brains slower but also increase the chances of heart problems. Plus, exercise helps our mood and can keep us from feeling too sad or anxious. So, it's important for us to make time for exercise, for the sake of the health of both our body and mind.

In recent years, a troubling trend has emerged: children are moving less and sitting more. This shift towards a sedentary lifestyle, dominated by screens, indoor activities, and reduced physical play, has serious health implications. Alarmingly, this lack of movement is now manifesting as a spike in obesity and diabetes rates among teenagers. Today's children are growing up in a digital era. With the advent of smartphones, tablets, and online gaming, outdoor play and physical activities often take a back seat. Additionally, academic pressures and increased homework loads limit the time children have for outdoor play. Urbanisation, with limited safe open spaces, further compounds the problem.

The consequences of this sedentary behaviour are dire. Childhood obesity has reached epidemic proportions in many countries. Obesity in childhood doesn't just affect one's physical appearance; it sets the stage for a range of health problems in adolescence and beyond. One of the most concerning outcomes is the rise in type 2 diabetes among teenagers, a condition that was once only prevalent among adults. Diabetes at such a young age increases the risk of long-term complications, including heart disease, kidney damage, and vision problems. Apart from the physical risks, being overweight or obese as a teenager has psychological implications. Adolescents often face bullying, self-esteem issues, and depression, further deterring them from participating in physical activities.

Addressing this crisis requires a multipronged approach. Schools, parents, and communities need to recognise the importance of physical activity for children. Efforts should be made to incorporate regular physical education classes in schools, encourage active transportation like walking or cycling, and promote sports and outdoor activities. Parents can also play a pivotal role by setting screen-time limits, being role models for an active lifestyle, and ensuring that family time includes physical activities. We'll look into solutions in more depth in Part 3.

You Are What You Eat

A healthy diet, combined with regular movement, boosts our mood and energy. Feeling good inside helps us face challenges head-on. Yet modern life, with its stresses, can lead us to poor food choices and less activity. I've had days when stress pushed me towards sweets and skipping my usual walk. It's not just about cravings; it's how our brains cope with stress and pressure, seeking quick comforts.

If we're not feeling healthy and fit, it's harder to go out into the world with confidence and purpose and to be an effective parent. It's harder to have the energy to do, well, a lot of anything. Yet the challenge is that our modern stressors often steer us away from being healthy and fit. I recall days when the stress of my research would see me reaching for sweets and chocolates, pouring an extra glass of wine, or indulging in a second serving of dinner. This wasn't mere indulgence; it was my brain's way of navigating stress, seeking instant gratification and the release of those feel-good chemicals.

Looking back, it's disturbing to think how unaware I was of how much food I was eating, how much weight I was gaining and how unfit I was becoming. Since my attention was on everything else, I didn't see what was happening to me. And, looking back, I see that I never felt full, even after eating a whole bowl of pasta and meat sauce and a big salad and drinking a glass of wine (and that's a whole lot of calories!). The more I ate, the more I wanted to eat, which is what happens with other types of addiction. Research has shown that the more we eat, drink, or take drugs, the less available the dopamine receptors are. This means that we must eat more, drink more, or take more drugs in order to feel the same level of pleasure. And this is why sugar is considered addictive. The more we have it, the more we want it.

Now, what does this mean for parenting and our children? Again, our children do as we do. They learn from us. Plus, our responsibilities as parents involve providing our children with nourishing and healthy food. Yet many processed and sugary food items don't offer what developing minds and bodies need to thrive. On top of this, consuming excessive amounts of sugar and processed food can have negative effects on a child's brain and cognitive development—and overall well-being. So, now we're going to take a closer look at how sugar and processed foods impact the brain, as well as provide simple and easy tips that you can implement in your family's life to develop healthy eating habits that last a lifetime.

Sugar and Processed Foods: What's the Problem?

WHO reports that about 39% of adults in the Western world are overweight. And this is due largely to high fat and high-sugar foods. While there are many factors at play here, including economic or financial situations, availability of types of foods, education, and more, the outcome is that in recent decades, we've become some of the unhealthiest generations to walk the planet.

Let's Start from the Beginning …

Sugar became a prominent part of the food chain in the late 1960s. It was used to replace fats, mask bitterness, and make food tastier. The result was that consumers would buy more of these foods, and the money would flow towards the companies providing them. While our ancestors obtained sugar from fruits and other whole foods, by the 1970s, we were consuming sugar at a much greater rate through the new ways our food was being processed. In the last 50–70 years, the worldwide population's consumption of sugar has tripled, largely due to hidden and refined sugars.

The health consequences of this shift are vast and tragic; they include increased rates of cardiovascular disease, diabetes, musculoskeletal disorders, and even cancer. Being overweight or obese as a child is also associated with a higher risk of being overweight or obese as an adult. And these negative health effects extend beyond the body; they also impact our minds.

Sugar Negatively Impacts the Physical Structure of the Brain

All this sugar *literally* changes the brain. I don't mean to use a fear-mongering approach; a treat every now and again is okay (remember, life is meant to be enjoyed!). However, research has shown that sugar has powerful addictive properties and impacts the brain in that it reduces impulse control, which explains why it's not easy to simply cut sugar high and dry.

Eating and drinking too much sugar and processed food leads to changes in the physical structure in the prefrontal cortex—the front part of the brain that helps us make good decisions. This area also helps regulate our emotions. So even though it might make us feel good in the short term, over time, it is changing our neuroplasticity. If we have too much during adolescence, sugar further affects cognition and our ability to make decisions and remember things.

Increased Sugar Consumption Causes Mood Changes

Increased consumption of sugar can also lead to an increased risk of mental health disorders like depression and anxiety. Furthermore, consuming sugar-laden processed foods can lead to imbalanced blood sugar levels, which can cause mood swings, irritability, and hyperactivity in children.

The truth is that there is a significant overlap between neural pathways that regulate our emotions and those that regulate the consumption of sugary foods. It's no wonder I loved stress-eating so much! It, physiologically, *may* help with stress—but not in the best way. Remember, eating sugar helps produce those feel-good hormones that we crave during times of stress. Yet this can lead to neural changes in emotional processing and, thus, behaviour. At the same time, rather than turning to sugary foods during times of stress, we may do better by looking for healthier ways to respond.

There's No Denying It: Sugar Is Addictive

Sugar can not only substitute for addictive drugs like nicotine, but can also appear to be *more* rewarding.

After my colleagues and I determined that sugar was just as addictive as nicotine, one of the most addictive drugs on the planet, we spent the next five years mapping, verifying, and replicating that discovery before confirming it and publishing a series of papers on the topic. The shocking finding was that sugar disrupts the brain pathways in the same way as alcohol and nicotine. Sugar consumption leads to an increase in the neurotransmitter that activates the receptors that nicotine binds to. It activates the reward system of the brain in the same way that abusing substances does. And, by the way, artificial sweeteners are as addictive as sugar, making them not exactly a viable or healthier replacement.

Of course, we can't go without mentioning that a sedentary lifestyle plays a part in the obesity epidemic. And while we talk about exercise in the next chapter, it's worth noting that moving our bodies activates those feel-good hormones as well and can provide a healthier alternative for combating stress (again, more on this later!).

So, let's break down everything further. What exactly is going on in the body when we eat sugar?

How Sugar Drives Us to Eat More

Sucrose, or table sugar, is the basic sugar we all know and love. Sucrose is made up of fructose and glucose—in other words, after we consume sucrose it is broken down into fructose and glucose. Every gram of glucose has the same number of calories as a gram of fructose, but these different types of sugar affect the body in completely different ways.

Glucose, also known as blood sugar, is the key source of energy for our bodies and brains. It's either used immediately when we consume it or stored as glycogen in the muscle cells or the liver for later use. Insulin, an important hormone, regulates blood sugar. It also lets the brain know that you're full and can stop eating.

Now, fructose is found naturally in fruit and vegetables, and it's *also* found in many processed foods. Yet fructose isn't what our brain and muscle prefer for energy. And, unfortunately, fructose has a major impact on the appetite centres of our brains.

This type of sugar blocks the release of hormones that help regulate our appetites, including those that tell us when we are full. Inevitably, consuming too much fructose can create some problems, such as overeating. In the long term, the part of the brain involved in appetite regulation weakens. We struggle to know when we are full, so we may eat more and more.

And get this: a chocolate bar contains around 50% sugar. (In contrast, an avocado contains only about 0.2% sugar.) The calories (or fructose) in this chocolate bar are most likely to be converted to fat. Whereas glucose gets shuttled into those muscle cells or used immediately, fructose doesn't function the same. And this is where the problem lies!

From there, we gain fat, which can be hard to lose, since the body only releases fat when we aren't getting energy from anywhere else. On top of this, stress (yes, stress again!) over time stimulates the growth of fat cells.

Another problem? Fructose (sugar) hides in many foods we eat. Sugar isn't only found in obvious places, like donuts, cupcakes, and soda. Sugar is embedded in the food chain. Data from the National Health and Nutrition Examination Survey and the US Department of Agriculture has shown that 75% of all foods and beverages contain sugar.

But this doesn't mean we're stuck with this way of eating. With a little awareness and a few simple tips, we can help our families and children thrive and avoid the mounting problems that sugar presents.

Chapter 10: LIFESTYLE FACTORS

The Five A's of Sugar

Ophthalmologist Dr James Muecke was the Australian of the Year in 2020. He has become an advocate of changing the nutrition and dietary guidelines to deal with the burgeoning impact of diabetes, which is now the leading cause of blindness among working-age adults in Australia. In fact, 98% of blindness and loss of vision is due to diabetes and is preventable.

Dr Muecke, a medical doctor of close to 40 years, has been at the frontline of treating these end-stage complications. Previously, he never thought it was his responsibility to have conversations about diet with his patients. Yet one day, Dr Muecke was sitting in the lecture hall listening to one of my talks about the addictive nature of sugar and our research on it. This was the first time it had clicked for him about how addictive sugar is.

He describes this as a powerful and pivotal moment in his career. Eventually, this led him to outline the five A's of sugar toxicity:

- **Addictive.** Sugar is as addictive as nicotine.
- **Alleviation.** We often use sugar to alleviate stress or to make us feel better when we're sad or down.
- **Accessibility.** Sugar is cheap, and it's everywhere, including the gas station, post offices, restaurants, stores, and more.
- **Addition.** Sugar is added to over 75% of processed foods and drinks.
- **Advertising.** Ads, like TV commercials and billboards, flood the market. Almost anywhere you look, you're seeing advertisements for a form of sugar.

It led Dr Muecke to make various changes in his own life and diet, such as cutting out and limiting added and refined sugar. Dr Muecke states that he was addicted to sugar, and it wasn't until he had a scan of his abdomen that he uncovered he had fatty liver disease.

Unwittingly, he was on a path towards prediabetes and type two diabetes. His biggest weakness? Ice cream. So, in January 2020, he went cold turkey, cutting sugar from his diet. But the withdrawal symptoms started on day one. He had headaches, irritability, fatigue, and cravings. After about three days, these symptoms began to subside. And he felt way better for it. No longer was he caught up in the cycle of sugar addiction. He had liberated himself.

You may not be aware that one of the most powerful marketing slogans ever created was the notion that breakfast is the most important meal of the day. This was popularised by John Harvey Kellogg in 1906 to market his newly invented breakfast cereal, Kellogg's Cornflakes. And to this day, there's not a shred of evidence supporting cereal as a nutritious breakfast.

The truth is breakfast cereals are packed with sugar. But with sugar, not all is what it seems!

Building Awareness of What We Are Eating

Food can be medicinal in nature—helping improve our health, nourish our bodies and minds, and even treat various diseases. But, in contrast, it can also be poison.

Within the intricate workings of the biochemistry of our bodies, fructose and alcohol are handled identically. Shocking, yes.

While our population is indeed eating too much and moving too little, the question many fail to ask is *why?* Why do we eat too much? Why do we exercise too little? Why has everything gone to hell in a handbasket over the last 40 or so years? Even though the food was plentiful before the 1980s, we didn't have the same problem. What's going on?

Dr Robert Lustig from the University of California, San Francisco, author of *Fat Chance* and *Metabolical*, spent the last 25 years piecing together the biochemistry and neuroendocrinology of the brain. He has investigated what is happening in the brain to cause this pandemic of not just obesity but also type two diabetes, hypertension, dyslipidaemia, cardiovascular disease, cancer, dementia, fatty liver disease, and even polycystic ovarian disease. He found that many of these diseases are associated with insulin resistance, which affects the mitochondria (the powerhouses of our cells). This means that our metabolic and energy systems are taking a serious hit. And this all comes down to what's in our food, which the food industry has drastically changed in the last few decades. Diseases that were somewhat rare are now affecting children when they didn't before. It's tragic when you think about it. And if you've watched documentaries like *Super Size Me*, you'll see how drastic these impacts can be, and how little time it takes for them to occur.

The easiest way to understand processed food is to take an apple as an example. Class 1 processing is an apple in its most simple form. Class 2 processing is apple slices in a bag, and Class 3 processing is applesauce made directly from crushed apples. Class 4 processing is that delicious and ever-tempting apple pie. Classes 1–3 are not the problem and do not cause disease. But Class 4 and its growing presence in our diets is where the problem happens.

Processing is when the food industry alters our food in some way. They might add ingredients, commonly sugar, emulsifiers, and/or additives, and they may strip it of ingredients and precious nutrients, like fibre. This processing alters the metabolic profile of the food, making it more of a "poison" than a "medicine".

Then, there's the problem, as discussed above: fructose drives us to eat more. So, what's happening here? Well, we aren't often eating more and more fruits and vegetables (which fructose could drive us to do and would likely have greater beneficial effects on our health). Instead, we're eating more and more processed foods. We're eating more and more foods containing empty calories, meaning they don't contain a very high nutritional value but do contain tons of calories. So, our population is getting bigger and bigger—and not because of high birth rates, but because of our increasing weight.

Time for Reflection

Remember, small steps can lead to significant changes. By reflecting on our choices and understanding the reasons behind them, we can make informed decisions that prioritise our well-being, allowing us to face life with energy, confidence, and purpose. Share your intentions with a trusted friend or family member, asking them to be your accountability partner. This shared journey can provide mutual encouragement and understanding.

Seeking Awareness. On a scale of 1–10, how aware are you of your food intake and physical activity? What steps can you take to be more aware of them?

Self-Inventory. Start by taking a moment to reflect on your day or the past week. Write down the foods you predominantly consumed and your physical activities, no matter how minor.

Linking Mood to Choices. Recall moments when you felt stressed, overwhelmed, or anxious. Did you seek comfort in specific foods or avoid certain activities during these times? Make a list.

Positive Choices. Now, think of a day when you felt at your best, both mentally and physically. What did you eat that day? Were you more active? Jot down these positive choices.

Analysing Patterns. Compare the choices you made on stressful days to those on your best days. Do you see any patterns? For instance, did stress lead to sugary snacks and less movement?

Once you've jotted down your reflections, take a look back. This exercise is not about criticism but awareness. Recognising our habits related to food and fitness helps us make better decisions in the future. Every small step towards a balanced lifestyle matters!

PART 3

Steps to Master Parenting in the Digital Age

Chapter 11

MANAGING STRESS

Rewire Your Overwhelmed Brain

In Brief

Stress is an inescapable part of human life, and it's particularly common among parents. The demands of parenthood, coupled with the daily pressures from work and social commitments, can easily push stress levels into the red zone. When our children act out or refuse to comply with basic tasks like going to school, eating, or sleeping, our stress levels can further spike. The problem is that stress doesn't just vanish into thin air; it lingers, affecting our physical health, emotional well-being, and even our cognitive functions. Therefore, it's crucial to understand the neuroscience behind stress and learn strategies for managing it effectively.

When something stressful happens, think of your brain like a smoke detector going off. As explained in Part 2, this "smoke detector" in your brain releases hormones, like adrenaline and cortisol, that prepare you to deal with the situation. The part of the brain that's in charge of this is called the amygdala, and it's the same area that helps us with our emotions and survival instincts.

This system is really good at helping us handle immediate dangers—like if you were to actually smell smoke or hear a loud noise in the middle of the night. The problem is, it's not so great at dealing with the ongoing, day-to-day stress most of us face. That's because it was designed for short bursts of action, not for long-term stress like juggling work, family, and other life pressures. This chapter will delve into how stress impacts the brain and offer techniques for "rewiring" to achieve a more balanced mental state.

Rewiring the Brain's Smoke Detector

We know the number one thing the brain likes is stress. In fact, the brain is faster at wiring itself in response to stress than it is in response to pleasure. Becoming aware of your stress reactions and knowing they sit within the smoke detector provides the opportunity to imagine what would happen if you were able to devise strategies to pause it. Akin to taking your brain to the gym and thinking of it more like a muscle.

Just as a "real" smoke detector sometimes goes off when there is no danger, our brain's smoke detector sometimes sends out false alarms; the perceived threat doesn't necessarily pose a risk to our survival. However, our body's physiological response is very much the same as if it did. But this is a good thing. We don't want our "real" smoke detector to go off only when the room is on fire. At this point, it's too late; we're done for. It's best the alarm starts ringing when there's even the faintest hint of smoke. In this way, we can escape and survive. Yet, of course, as with a smoke detector, there are going to be a few false alarms when it comes to our stress response. Remembering that the room isn't necessarily on fire every time the smoke detector goes off can help us create space between the stressor and response. It can also allow us to adjust before situations become dire—like putting out the grease fire on the stove or opening a window to allow proper ventilation to let cooking smoke out.

The smoke detector is something I like to compare to a musical score, which can be fast paced without any breaks in the notes, like a march. When the brain is being trained like a muscle, we spread out the number of false alarms—or, as in a music score, we add pauses. This concept of understanding the power you have to put in pauses with brain training is at the epicentre of real change in the brain—and your life. Fortunately, we can pause the smoke detector. We do have some control over our response.

CASE STUDY – Helen

Helen is a prime example of how we might control our response. Helen was a mother of two, a six-year-old daughter named Amy and a three-year-old son named Max. She loved them dearly but found herself often getting stressed out when they refused to listen or cooperate. One day, Amy woke up crying and refused to go to school. Helen's first instinct was to become frustrated and raise her voice, but this time she stopped herself and took a deep breath. She knew she needed to handle the situation differently.

Helen had been training her brain using exercises daily to pause the smoke detector and bring it to her conscious awareness. Instead of forcing Amy to go to school, she sat down and talked to her. She asked Amy what was wrong and listened patiently as her daughter expressed her fears about a bully in class. Helen acknowledged Amy's feelings and talked about ways to handle the situation, such as talking to the teacher or asking for help from a friend. Amy felt heard and understood, and together they came up with a plan.

Later that day, at the supermarket, Max started throwing a tantrum when his mum, Helen, wouldn't buy him chocolate. In the past, she might have given in to avoid the embarrassment of a public meltdown. But this time, she remembered her smoke detector, took a deep breath, pushed her shoulders back and took a different approach. She knelt to be at his eye level, adopted a calm tone of voice, and gently talked to Max, asking him what was wrong. In time Max chose a healthy snack instead. Max stopped crying and hugged his mum, and a lesson was learned for both of them. Max learned his mum did not give in easily; Helen learned a new way to respond when she was stressed out and in a difficult situation and highly charged moment and time of the day. Like learning a new language, Helen and Max are learning a new way to get along and dance together.

Through these experiences, Helen realised that being aware of her smoke detector and training her brain daily had made a significant difference. She was able to respond to stress in a more positive and constructive way. She knew that she would still face challenges, but with her newly trained smoke detector in her brain, she felt more equipped to handle them and build a deeper relationship with her children.

Often, with stress, our neural circuits (the smoke detector) quickly refer to the memory banks of the parenting model learned from our parents. As a result, this can lead us to react quickly (and, sometimes, badly). However, it's important to give ourselves some grace here.

At the end of the day, we have the power to become aware of the smoke detector, learn tools to pause it, to turn millisecond reaction times into one-second responses. This is enough time to engage other parts of the brain and be able to take a wider view of the situation. In essence, we are training our brains to rewire stress *reactions* into stress *responses*.

If this was easy, everyone would be doing it. Yes, it takes some effort and practice, but when we experience the positive effects for ourselves and see the positive influence it has on our children's development and health, it's entirely worthwhile. What was wired into the brain over centuries and millions of years does not get rewired overnight—unfortunately.

As parents, we are constantly making decisions that can have a significant impact on our children's lives. ***When we understand the way our brain responds to stress, we can:***

1. learn to retrain our innate stress reactions—the smoke detector
2. work on managing and then preventing the impact from chronic stress, which takes a toll on our mental and physical health
3. respond more effectively, nurturing our children's brain health and development.

It's important to note that a smoke detector doesn't define us or our identity; it is simply a neural circuit and an important part of the brain. We have the power to hit the pause button on the smoke detector and take control of the steering wheel, which directs our lives and our stress reactions. Acknowledging that we want to make a change is the first step; then it's about taking action to train our brain towards where we want to go and who we want to be. Here are some quick ways to alter the stress response.

Quick Ways to Alter Your Stress Response

1. **Changing Your Body Position.** You can put your arms up and stretch towards the sky or place your hands on your hips with your shoulders back. These body postures demonstrate a dominant position in the animal kingdom and promote positive neural chemicals, dimming that

stress response. The main outcome is feeling more confident and driving in a one-second pause to the situation.

2. **Take in a Panoramic View.** Physically put your phone, tablet, or any other distracting device down. Position yourself so that you have a wide, panoramic view of the outside world. This could be by a window, on a balcony, or in an open outdoor area. Make sure your field of vision captures as much of the environment as possible.

3. **Performing a Tracing Exercise or Doodle.** This means calmly tracing a drawing or picture with a fine pen or pencil as carefully as you can. Try to go slow and focus on precision. Within a few minutes, you should feel more at peace and less on edge.

4. **Using a Miggi Jar (short for amygdala or the smoke detector jar).** This means making a jar filled with water and glitter. You can add beads or anything else to the jar that represents your worries or stressors. Put some feathers on top of the jar and when you feel stressed, shake the jar, hold the jar at arm's distance and then blow into the feathers. As the beads and glitter settle, the water will be transparent again, and you may feel yourself calming down. Do this as many times as you need to. (This can also be a helpful way to help your children deal with stress!).

5. **Move Your Body.** Exercise is a wonderful way to induce the release of feel-good hormones, helping to dampen the stress response. You may choose to stretch, walk, swim, bike, or simply shake your body a bit to release those stressful feelings.

With the above exercises, we can pause the smoke detector and calm our bodies and minds. When we use these activities, we further

Chapter 11: MANAGING STRESS

acknowledge that the brain can be trained like a muscle. We aren't stuck in our ways; we *can* change!

Train Your Brain to Respond *Rather than* React *to Stress*

One of the easiest things we can do to nurture our brain health and our children's brain development when stress arises is take time to pause.

Recognise you're stressed. Take a step back and breathe for a moment. Then respond thoughtfully, rather than react. Each time you do this, your brain learns. Plus, taking a deep breath naturally lowers our physiological response to stress and lowers our heart rate. If the situation allows, we can also decide to take a "brain break". This may mean using the exercises described above.

As many other parenting books suggest, fostering good time-management skills can further help us as parents give ourselves space to avoid stress. This may mean planning an extra 15 or 30 minutes when getting our family ready to go to an event or gathering, as this may account for the unexpected, like temper tantrums or refusals from our children to get ready.

Lastly, when we feel stressed or anxious about a deadline or interview, or when someone has asked us something difficult and we feel those butterflies in our stomachs, we can step back and take a few moments to practise in front of the mirror. Say how we are feeling. This will help change the pace of the smoke detector's ability to send off false alarms. As with learning a new language, we must start small and build—but every little bit counts!

In my previous book, called *Smashing Mindset*, I included some principles to help us train the brain to manage stress effectively and positively and help us and our families lead healthier and happier lives. To effectively manage stress, it's crucial to first understand its neurological impacts on your brain's wiring, as this can influence your behaviour and overall well-being. Armed with this knowledge, you can then master specific techniques focused on retraining your brain to approach stress with a more measured, thoughtful response as opposed to an automatic, reactionary one. Without stress awareness it is not possible to make any change. The smoke detector and its neural connections across the brain and throughout our body are very old and hardwired.

Everyone's reaction to stress is varied because of the imprints left in the brain from early life experiences. This means the sensitivity of their smoke detector varies. For example, a person who explodes quickly has a brain with a highly sensitive smoke detector. Think of a smoke detector having a range from 1 to 100, with a score of 90 meaning the person explodes when someone cuts them off in traffic to a score of 10 for someone who lets the person pass in front of them. The smoke detector is the brain's threat-detection centre that is wired by early childhood experiences, and in some cases inherited from parents and grandparents and further back in time. This circuit is activated in milliseconds and leads to fast responses when people feel under threat—responses such as shutting down, lashing out, getting into a rage, isolating at home, running away, drinking alcohol, taking drugs, overeating, gambling, becoming addicted to adult material. There are many stress reactions that create havoc in people's lives because they are outside their awareness.

What You Can Do for Yourself to Feel Good

What are you paying attention to? For example, if you are always scanning the news, notice that you may begin to see the world as dangerous. In contrast, if you are in the garden, walking on the beach and in nature, and not paying attention to the news, notice how you feel about the world. Because the brain is scanning our environments in the search of threats, we pay attention more to negative events than positive ones. Therefore news, social media, and movies stream scary, frightening views of the world to garner your brain's attention. This activates the brain's alarm systems, and, in a spiral, your brain seeks further confirmation of the bias and beliefs now created in your mind.

Imagine instead someone handing you a crystal ball that plays the film of your future life: one version where you stay the same, and the other after you spend the next ten seconds doing something you have never done before. If your daily routine is as tepid as the water from your tap, it's time to dip your toes—nay, your whole body—into a cold bath. I mean *literally*; extreme temperatures can shake up our senses while calming us down, making it worth our while to plunge out of our comfort zones.

I was standing in the brace position taking my daily shower and, instead of enjoying my usual hot steamy shower, I turned the tap to cold and pointed it towards my toes, immersing them in cold water. The next day, I did this for 20 seconds, and over time, I let the cold water creep up to my knees, spine, and back of my neck. Little did I know, the ten seconds it took to decide to take a chance and say yes to a cold shower would be the step towards disrupting my entrenched belief about cold showers. I had been the queen of the steamy hot shower, and always sitting on the edge of a pool in summer. This one belief I held about the cold was a habit hardwired into my brain. Cold showers made me step back and reassess my beliefs.

The advent of technology has led to an opening up of conversations about the mind, and there are many examples of people who have been experimenting with the limitations (or not) of the brain's capacity for change. Wim Hof, for example, is a Dutch motivational speaker known as "The Iceman", who trained himself to tolerate swimming in icy conditions. The Wim Hof Method, like other examples of neuroplasticity, shows how the brain can be trained like a muscle. Thousands of people around the world have since used Wim Hof's simple brain-training techniques.

When we repeat the same activities for years, the brain no longer has to think about what to do next—it's on automatic pilot. This is great when we have mundane tasks to achieve every day. But doing something new every day is a great way to train the brain to feel comfortable with discomfort, and you don't need to jump into an icy lake or take extreme measures. After all, there is always a first time for everything. In under ten seconds, Wim Hof's life was forever changed when he decided to jump into an icy lake. The icy lake helped him feel alive again for the first time in years, and the cold became his new friend. Have you ever wondered what change might be in your future by thinking, feeling, or doing something new for ten seconds?

Wim worked alongside scientists to prove that his breathing exercises and cold exposure changed his physiology and brain's ability to respond to stress. Along the way he discovered how to access the deepest parts of his brain that interact with the autonomic nervous system (the part of the brain that is outside of our conscious control). Dr Vaibhav Diwadkar, professor of psychiatry and behavioural neuroscience, and his Wayne State University colleague Dr Otto Muzik, professor of paediatrics and radiology, were the first researchers to take an image of Wim Hof's brain. Wim Hof was accidentally rewriting the medical and scientific textbooks. What we had thought of as the autonomic nervous system—outside our conscious control—*can* be taken into conscious control and trained like a muscle. The Wim Hof Method is now used by thousands of people. It has been published in top scientific journals, such as the *Proceedings of the National Academy of Science*, and shows the brain is the most powerful untrained machine we each have access to.

Below are some more tips for making yourself feel good.

Try a Cold Shower

A cold shower in the morning offers ample benefits. It doesn't just boost the immune system, as popularised by the Wim Hof study. This simple change can calm anxiety and stress, as well as potentially reduce symptoms of depression. It's thought that taking a cold shower works by increasing blood circulation throughout the body and releasing feel-good chemicals called endorphins, which can calm us and promote well-being. The perfect way to start your day on the right foot! Cold immersion is a great way to turn the tap on and access neuroplasticity.

Here's how to get started with a cold shower in the morning:

- If you're new to cold showers, try starting with your normal water temperature and then turning the tap to cold for 10 seconds at the end of your shower.
- The key is that you want it to feel uncomfortably cold.
- Eventually, try to stay under the cold water for at least 30 seconds or a couple of minutes.
- Ideally, you can also finish with the cold water. You may also find as you get used to the cold temperature that you can go colder over time.

Breathe Mother Father

Wim Hof was asked on a podcast, "If you could have a billboard what would be one thing you would write on it?" He said, "Breathe Mother Father."

Deep breathing, also called "breathwork", can be an invaluable tool for dampening the stress response. This can be performed in the morning and throughout your day, as needed. The key is to take deep, slow, and controlled inhales and exhales. Doing this activates the amygdala and the parasympathetic nervous system (the rest-and-digest system). In turn, breathwork offers a powerful way that we can re-centre ourselves during stressful or anxious times. Many find breathwork to be quite useful before bedtime, and it's been shown to help improve sleep (which, as we know, can also help reduce stress). This would make a huge difference to improving the brain health for parents and children. The effect would then be passed on for generations to come as a better way to mitigate and handle stress. Imagine putting into your day some deep breaths and knowing that you are retraining the parts of the brain circuitry that process stress, fear and anxiety. It is a simple tool that has been used for centuries.

A popular breathwork method for reducing anxiety and stress is called the 4–7–8 method, which involves the following.

1. Close your eyes.
2. Take a deep, full-belly breath through your nose to the count of 4.
3. Pause for the count of 7.
4. Exhale through your mouth to the count of 8.
5. Repeat this cycle 3–5 times or until you feel calm.

A more simplified version of stress-relieving breathwork involves taking a deep breath in through your nose to the count of 4, pausing for 4, exhaling through your mouth for 4, pausing again for 4, and so on. If this is easier to remember, especially when stressed, do this activity instead.

Start Your Morning with a Panoramic View

The very first thing you can do that will help lower stress immediately at the start of your day is to *look outside at the horizon instead of your phone.* We know that using our eyes and taking in a panoramic view activates the visual nervous system and deactivates the stress system. There is a direct connection between the visual circuits and circuits that regulate stress, like the amygdala. We can make a conscious choice to direct the brain's attention to regulate stress.

While stress isn't avoidable, we can learn to calm it and learn what our stress response is trying to tell us (it's not all bad!). Maybe we need to change what we are doing or view the situation from a new perspective. Often the morning routine is the only time many parents have to themselves, even if it's only a few minutes. This time to ourselves can give us a bit of space to prime our brain to react calmly as we approach situations throughout the day.

To perform a proper morning routine, we might need to manage our time better—plan ahead, set our alarms slightly earlier, or even simply get up as soon as our alarms go off (rather than hitting snooze). For some parents or carers, the best time to do something for themselves might even be after school or daycare drop-off. Determine what works best for your schedule and pencil it into your calendar! Now, what should you do to set the stage for a successful (and less stressful) day?

Practise Gratitude

Some may also find it helpful to perform a quick gratitude practice in the morning. This can set the stage for the rest of your day. And this is simple. Set a time limit (such as five minutes) to list out everything you are grateful for in your life. If you don't have five minutes, try to list just three things. These can be big or small. The key is to truly feel those feelings of appreciation. Research shows that practising gratitude regularly can have profound effects on brain health, including reducing perceived stress and symptoms of depression. A further suggestion for practising gratitude is to reflect on the individuals who are thankful for the assistance or care you've provided them. What goes around comes around!

Try Nature Bathing

Making time for nature (or even with your kids) can have significant benefits on stress levels and brain health. In Japan, there is a practice called forest bathing or *shinrin-yoku*, which literally translates to "forest bath"—that is the practice of "bathing" oneself in nature with the intention of receiving therapeutic benefits. The research suggests there are some benefits from surrounding ourselves with trees and immersing ourselves in the natural environment. Numerous studies have shown the benefits of forest bathing on our mental and physical health, although the quality of the studies varies from low to moderate. The main benefit that I derive from being in nature is the ease of engaging the senses and being fully present in its beauty. The sounds of birds chirping, the scent of fresh trees, the feel of the cool breeze on my skin. You may even want to completely immerse your senses by identifying one thing you feel, one thing you smell, one thing you hear, and so on. Now, incorporating this into your routine doesn't have to be complex. It can be as simple as taking a leisurely walk in a nearby park or forest, breathing in the fresh air, and fully embracing the natural surroundings.

For an instant release from stress, if you are able to, take a leap into the ocean. Feel the embrace of the saltwater as it gently strips away your worries, leaving a trail of tranquillity that touches your soul, grounding you with the earth's serene simplicity. Being in nature is powerful and can help ground us and connect us with the natural world. Simply disconnect from technology, let go of distractions, and allow nature to rejuvenate your mind!

Move Your Body

If you have the time, exercise is also a great stress remedy. While we discuss this more in later chapters, your morning routine may consist of some form of exercise, such as walking, stretching, or even a quick strength or cardio workout. Exercise has the powerful ability to produce endorphins, which counterbalance those stress hormones.

Try Healthier Food

Lastly, a good morning routine to fight stress also includes healthy food. Again, we will talk more about nutrition and its role in brain health and development later. However, since the stress hormone cortisol is naturally at its peak in the morning, having a healthy and balanced food with protein, carbohydrates, and fats within the first few hours of awakening may help control stress, especially that associated with morning anxiety.

Mindfulness: A Double-Edged Sword for Young Minds?

Mindfulness, with its roots in ancient traditions, has seen a resurgence in modern psychological circles. Heralded for its potential to decrease stress and anchor individuals to the present moment, it's being prescribed more frequently as a panacea for today's fast-paced world. Yet a growing chorus of experts is raising cautionary flags, particularly when it comes to its applicability for children and adolescents. Contrary to the prevailing notion, surprisingly, mindfulness and meditation might not be beneficial for children, having either no effect or even leading to increased mental health issues.

It's time we mind the hype and shake the idea that there is a "one-size-fits-all" approach for mental health across all ages. There is no evidence indicating the benefit of mindfulness for children or adolescents. For those children more susceptible to mental health problems, mindfulness and meditation put them at an even greater risk of succumbing to anxiety or depressive disorders.

So, what does this all mean? Well, we shouldn't necessarily encourage meditation or mindfulness practices in our children's daily lives.

It's about Doing the Best We Can

Armed with this knowledge, you can master specific techniques focused on retraining your brain to approach stress with a more measured, thoughtful response, as opposed to an automatic, reactionary one. Food and diet play a pivotal role in managing stress; minimising consumption of foods and beverages high in sugar and fats can alleviate stress-induced biochemical changes. Physical activity serves as another potent tool for stress control, offering not only an immediate release but also long-term benefits for mental health. And don't underestimate the power of sleep, attention and connection as antidotes to stress; fostering awareness and nurturing social bonds can go a long way in achieving a balanced, less stressful life.

No one is perfect, so we can't expect ourselves to be either. While we can limit adverse effects on brain health, it's near impossible to eliminate them. Thus, it's about doing the best we can. For example, we could ensure a proper serve and return relationship, emphasise eating whole foods over processed foods, and continually work on bettering ourselves to provide good examples for our children. We can tackle issues, as a family, without yelling and in a calm, loving, and safe manner. We can teach our children about alcohol and demonstrate its use in a safe and healthy way. We can set proper bedtimes and bath times. We can create a family dynamic of safety, security, and love, including eating dinner together, reading bedtime stories together, offering a place to express ourselves freely and feel all the emotions that come with the human experience, and limiting times around digital devices (such as no phones or screens in the bedroom). All these things, along with an appropriate amount of discipline, contribute to a safe and calm family environment that nurtures our and our children's brain health and fosters deep connections within our family unit.

As we navigate the intricate balance of raising well-adjusted children in an increasingly complex world, it's clear that the pursuit of quality time becomes the golden thread that ties all these efforts together. Quality time isn't just about the hours we spend, but how we spend them. It's the safe and nurturing environment we create, the choices we make, and the values we instil that set the stage for deeper, more meaningful interactions. The choices we make in these moments don't just build stronger family bonds; they build healthier, happier brains as well.

Time for Reflection

These exercises aim to help you reflect on various daily habits and actions that can positively influence your mental well-being. By actively participating in these activities, you can promote brain health and reduce stress for both you and future generations.

Cold Shower Challenge
- Think about your current shower habits. Have you ever tried a cold shower?
- Challenge yourself: For the next week, end your regular shower with 10 seconds of cold water. Gradually increase the time as you become more comfortable.

Breathing Exercises
- Try the 4–7–8 method or the simplified version. How did you feel afterward?
- Set a reminder to engage in this breathing exercise at least once daily.

Morning Routine
- Spend five minutes every morning looking out of a window, balcony, or outdoor space before looking at your phone. Focus on the horizon and take in the view.
- Reflect on how this practice affects your mood and stress levels throughout the day.

Nature Immersion
- Plan a "forest bath" or nature walk this week. If possible, bring along family or friends.
- Reflect on your senses during this time. What did you hear, smell, see, and feel?

As you engage in these activities, remember that every small action can ripple into a significant impact on how you feel. It's not just about reducing stress but fostering a deeper connection with yourself and the world around you. Embrace these practices, and let them guide you towards a brighter, more balanced life.

Chapter 12

RE-PARENTING

Invaluable Benefits for Children and Future Generations

In Brief

In 2009, I found myself navigating the highway called the Interstate 80 near Berkeley, California—the pulsing vein of Silicon Valley. At that moment, I felt like a plate juggler in the middle of a high-stakes performance, attempting to keep a multitude of responsibilities airborne. And while I'd like to say I kept all those plates perfectly balanced, the reality was that many of them came crashing down. But at that moment, driving my Honda CRV with the engine's steady purr mirroring my inner resolve, none of that seemed to matter. As the sun dipped below the horizon, casting its golden farewell over the iconic Golden Gate Bridge and the rolling Berkeley hills, I felt invincible, as if I were living my best life. Sharing this ephemeral moment with me was Ava, in the passenger's seat. She wasn't just any companion; she was a dedicated teacher from my children's after-school program, a woman who had invested years in nurturing their young minds.

Ava was a remarkable woman, a beacon of resilience and strength in a society that can often feel like an insurmountable mountain, especially for those who arrive without the armour of education and opportunity. She had left her native country, trading the familiarity of home for the uncertainty of a new life, all in the hope of offering her children greater prospects. Raised in a large family, Ava would recount stories of her shoeless treks to school, her empty stomach a testament to the hardships she faced.

Yet, her story, so compelling and full of struggle, is almost too vast to be confined to mere words on a page. Despite battling chronic health conditions that would have deterred most, Ava not only moved to Berkeley but also carved out a career as a Montessori teacher, a role she held for an astonishing two decades. This wasn't just a job for her; it was a lifeline, providing her children with an education and a community of support that became an extended family. There were times when Ava, with a nine-month-old baby in tow, would spend her nights in a homeless shelter or even a park, only to show up at school the next day, steadfast in her commitment to care for our children.

Ava looks out of the window of the car for a moment. "You know," she says, breaking the comfortable silence, "I've always believed that people can be incredibly strong and resilient, given the chance." I turn to her and say, "Ava, not many of my friends would be able to achieve what you have. We grew up with opportunities, housing, food, and education." Ava's daughter achieved a milestone that marked a turning point in their family's history: she became the first to earn a full scholarship to the prestigious University of California. The intricacies of this incredible journey could fill an entire book, each page a testament to the battles fought and the obstacles overcome.

By achieving this, Ava had shattered generational cycles of disadvantage, forging a new path not just for herself but for her family's future as well. The role of the community in uplifting and supporting her and her family was indispensable. Together, they had transformed not just one life, but set in motion a ripple effect that promised to impact generations to come.

"Imagine if we could provide that kind of support to everyone," Ava muses. "How much untapped potential could we unleash?" So, if you ever find yourself wondering, *Is it really possible to break free from*

Chapter 12: RE-PARENTING

the shackles of multigenerational trauma, stress, disadvantage, or flawed parenting? pause and think of Ava.

She isn't a mere figment of imagination; she embodies the resilience that resides in each of us. Through such concerted efforts, cycles of disadvantage are not just disrupted; they're metamorphosed into a continuum of progress that benefits future generations.

The saying goes, "It takes a village to raise a child", but there's another, less-discussed aspect to this proverb. A child who doesn't receive the necessary support early on can, in essence, set the entire community ablaze. We live in a time where mental health struggles are escalating, and where screen time is often supplanting genuine human interaction. The consequences of neglecting our collective responsibility towards our young can reverberate through communities, manifesting as widespread issues that we all end up grappling with. Hence, the investment in each child is not just an act of compassion but a crucial societal imperative.

We have the tools to understand and acknowledge the root of our pain and begin to heal using neuroplasticity, the remarkable ability of the brain to change forever. How do we gain the courage to see ourselves in a new light in the dark cave of our collective humanity? One step is to see how far we have come from the darkness of the past using the latest in DNA sequencing technology.

Let's consider that it was only in 2012 that it became possible for anyone to have a portion of their DNA tested for around $150. (To compare, in December 2001, the first whole human genome was sequenced for US$2.7 billion). I remember the day when I had the opportunity to find out where I came from. Family history is carried in genes that can be decoded using DNA sequencing machines. The DNA read out is the blueprint, or instruction manual, of our physical body, brain and mind. As a reminder, genes are made up of DNA. Some genes act as instructions to make molecules called proteins.

The instruction manual we inherit provides a starting point from our immediate past that includes parents, grandparents, and great-grandparents, and our distant ancestors dating back to the beginning of the universe, such as bacteria, mushrooms, and many plant and animal species—it was not fun to learn I was an evolving mushroom, but on reflection it made a lot of sense.

The cardboard box arrived from 23andMe, the DNA testing service that offered genetics-based ancestry reports. It was exciting to open the package, to take out a tube and know that by simply spitting into it, eventually parts of my family history would be revealed. Fast forward ten years and there are now millions of people using the 23andMe platform, and other genetic testing companies such as Ancestry.com. I've now met second and third cousins from around the world. But the most startling family history revelation came while I was watching *Outlander*, the time-travelling romance drama partly set in the Scottish Highlands.

One of the main clans on the series, the Frasers, have many headstones at the site of the Battle of Culloden, which took place in 1746. It turns out that I'm a descendent of Alexander Fraser, who set sail from Scotland to Australia in the 1800s. Just think about it: in 1746, the average life expectancy was 43 years compared to 73 years now. The staple highlander diet then consisted of oatmeal porridge, barley cakes, and some vegetables and dairy basics, and most of a person's day was spent collecting and preparing that food. Many people died in infancy or early childhood, or from infectious diseases.

Germ theory only became widely accepted in the early 20th century. Before then, medical professionals did not sterilise equipment or wash their hands before treating a wound or performing surgery. And over in the judicial system, the standard method of punishing criminals was to hang them, their execution made a public spectacle to act as a deterrent. It can be an interesting exercise to reflect on where we came from and how far we've come. The inventions of antibiotics, vaccines, refrigeration, electricity, and transportation have made life much easier. We have food delivered by bike, cashless payments, and phones that are handheld computers. Yes, we have large problems to solve, but imagine the collective, inventive power we humans have at our disposal to develop ingenious solutions that will renew our resources.

You don't have to get your DNA sequenced right at this moment, but when notifications pop up on the phone or when your food comes out of the oven, know there were generations of people whose creativity and willpower made those things happen. We're far mightier than we realise, and focusing on our strengths—and not sweating the small stuff—may be the elixir that helps us make it better for the next generation. For me, thinking about history has been a great way to relieve modern-day stresses such as emails and bad news. I calm myself thinking about the things that I can be grateful for now, and how much worse life would be if I were to travel back in time. We can all do this exercise by making the simple choice to start each day with at least three things we're grateful for, such as our friends, health, and home. We can do this rather than remind ourselves of the things that made us stressed out yesterday. Why do we stay stuck in the past and find it enormously difficult to see our resilience and strength; why don't we move forward to create brighter futures for the generations coming behind us?

How to Break the Transmission of ACEs from One Person to the Next

Dr Anda, one of the pioneers of the Adverse Childhood Experiences (ACEs) study, says our society needs to understand that ACEs happen because of person-to-person transmission. Adversity leading from trauma is like a virus. In medicine, we talk about preventing vertical transmission of lots of viruses, usually from mother to child during pregnancy. We would ask questions like, "Where are the most important places where this is being transmitted, and what can we do to stop them?" One of the important initial places of transmission is between caregivers or parents and the child. The way I treat another person stimulates their five senses, and they take it in and process it to decide whether my behaviour is a threat or not. Many of us have the vague notion that our response to others is just psychological, just something in the ethers. We now know we create a *biologic response* in every person we interact with, whether it's through the way our face looks, or our smile, or the way we smell or whatever. We are constantly transmitting experiences to each other. Just what experiences are these, and how do they influence human development and behaviour and well-being, not only in childhood, but even into adult life and older adult life?

We are not born a blank slate. The day we are born was a genetic lottery, and the chances of being born to loving parents were slim. From the beginning, there is about a 60% chance of inheriting genes that leave us susceptible to developing mental health disorders. Coupled to that is the fact ACEs are the leading factors steering an individual's susceptibility to developing mental health disorders. It is astounding that there are so few parenting manuals that have instructions for reducing the transmission of ACEs and the ingredients necessary to assist parents to raise happy, healthy, and strong children. Many of us tend to parent our children like our parents did or by doing the exact opposite. Have you wondered how we can inform new and old parents about ACEs in an informed, rigorous and life-affirming way? What would be included in this manual, and what would it look like?

Dr Catherine Lebel (Canada Research Chair in Pediatric Imaging) has done vast research on how infants exposed to higher prenatal stress are more likely to have a smaller amygdala (the fear centre of the brain). So, what's the big deal? Well, a smaller amygdala is commonly associated with behavioural disorders and higher anxiety in children, indicating the undeniable role that the parent's stress levels play in a child's development and how early childhood experiences can shape the brain.

When looking at children who have suffered trauma early on, we also see that the centres of the brain responsible for emotional regulation have reduced activity. Inevitably, if not addressed, this can lead to detrimental issues later in life, impacting a person's ability to socialise, interact with others, learn new things, feel a sense of belonging, and experience good overall health and happiness. And again, parenting isn't easy. It's stressful in multiple ways. But we have immense control here. As parents, we can learn to balance the stressors (which will happen) with practising compassion for ourselves and our children. And as mentioned previously, there are many simple ways to do this, which we'll explore shortly.

Interestingly, having a supportive partner or community throughout pregnancy can sometimes mitigate stress and lower its impact on the child's development. This makes sense; as humans, we are very social species. Our ancestors lived in tribes for this very reason; it helped them cope with the adverse events that everyone faces as they journey through life. Additionally, nutrition may play an invaluable role in dampening the stress response, which we'll discuss in more detail in the chapter on food.

Luckily, some youth organisations are starting to take note of the importance of ACEs and the impact they have, and how they affect a person's life. The organisations Pale Blue and HEERO™ (Helping Everyone/Each Other Reach Out) have made this a key piece in helping youth involved in child welfare–related care and juvenile justice systems. Here is a story about some young people who HEERO™ was striving to help. Spoiler: their story has a happy ending; they eventually became "peer navigators" in the program after overcoming a lot of adversity in their lives.

Unfortunately, their foster care placements turned out to be abusive situations. When this was realised, they were moved about many times, never truly experiencing a sense of permanency. However, their stories began long before they entered the child welfare system. Previous generations had also been placed in foster care. There was undeniable trauma in these previous generations, and many of these relatives had harboured pain and trauma throughout their entire lives. And this pain was passed down through generations.

Yet Pale Blue and HEERO™ offered these young people the ability to understand the ACEs everyone in their family had experienced. They began to understand that each person was simply doing the best they could with the resources they had. From there, they were able to shift blame and develop a close relationship with the family members who wanted to see them. This understanding allowed them a pathway towards healing and moving forward in a healthy and constructive way, especially after leaving the child welfare system and the lack of a support network. HEERO™ provides a network for individuals and supports them through this transitional process, helping them form close bonds for life. Many young people credit this as a focal point in changing the direction of their lives. Today, peer navigators play a vital role in helping other young people make the transition from the child welfare system to productive, healthy, and successful adulthood. The peer navigators teach the importance of courage, worthiness, and vulnerability, as well as the foundations of ACEs and the impact this has on the brain.

While this is just one example of how recent findings in neuroscience are being applied to help individuals thrive and move past trauma, there are countless more. The goal within these pages is to help you as a parent or carer to learn how to prevent ACEs to the best of your capabilities. Because stress simply won't go away. Stress causes the brain to adapt, altering its structure and keeping us stressed, and echoing through generations down the line. Despite this, neuroplasticity supports the idea that we aren't stuck. While the brain isn't born a blank slate, and everyone has different starting points, we can positively impact the structure of the brain, especially early on in our children when the brain is highly plastic. Before we delve into methods for mitigating stress in your life—and subsequently reducing its impact on your children—it's essential to explore how stress physically manifests itself.

Chapter 12: RE-PARENTING

Addiction serves as a compelling example. Be it to prescription drugs, gambling, work, food, sex, pornography, alcohol, or nicotine, addiction is often the body's misguided attempt to self-medicate stress. The relationship between stress and addiction is not merely emotional or psychological; it's a complex interplay involving both brain and body. ***In essence, addiction can be seen as a form of stress medication.*** So why do we become addicted? The answer often lies in our struggle to manage stress effectively, leading us to seek relief in unhealthy behaviours.

From a neuroscientific perspective, addiction can be understood as a coping mechanism for unaddressed or unresolved ACEs. The brain is wired to seek pleasure and avoid pain. When faced with stressors, especially those originating from traumatic experiences in childhood, neurochemical systems in the brain are activated to manage that stress. One of the most well-known systems involved is the reward circuitry, including the neurotransmitter dopamine. This circuit is activated not only during pleasurable experiences but also in response to substances or behaviours that are addictive.

In individuals with ACEs, the brain's stress response system, involving the amygdala, hypothalamus, and pituitary gland, may be chronically activated. This chronic activation can lead to what is known as "allostatic load", essentially "wear and tear" on the body and brain. Over time, this can result in dysregulation of the brain's reward system. People in this situation can become more susceptible to addictive behaviours to self-regulate or self-medicate their stress and emotional pain.

Additionally, ACEs are often linked to changes in brain structures and functions, including reduced volume in the prefrontal cortex—the area responsible for executive functions like decision-making, impulse control, and emotional regulation. This impairment can lead to difficulty in making healthy choices and increased susceptibility to addictive behaviours.

Hence, when the addictive substance or behaviour triggers the release of dopamine, it temporarily masks or numbs the emotional pain stemming from ACEs, providing a short-term solution for a deep-rooted problem. The brain then associates the addictive behaviour with relief from stress, reinforcing the cycle of addiction as a form of self-medication.

Understanding this neuroscientific basis of addiction in the context of ACEs is crucial for developing targeted interventions and treatment plans. Addressing the underlying childhood traumas and retraining the brain's stress response system can be key steps in breaking the cycle of addiction.

Neuroplasticity and Overcoming Echoes from the Past

Neuroplasticity offers a ray of hope in overcoming the long-lasting effects of stress and ACEs. Once thought to be a rigid structure, the brain is now understood to be highly malleable, capable of forming new neural connections and pathways throughout life. This adaptability means that the negative neurobiological imprints left by stress and ACEs aren't set in stone. In the previous chapter we outlined the strategies for individuals to rewire their brain's stress response systems. By doing so, they can reduce the impact of chronic stress on the brain, which is often overstimulated in people with ACEs and contributes to heightened stress and emotional reactivity. Simultaneously, they can strengthen the prefrontal cortex, enhancing executive functions like decision-making and impulse control that are crucial for emotional regulation.

Essentially, neuroplasticity allows for the brain's architecture to be remodelled, providing a biological basis for recovery and resilience. This makes the concept pivotal for treatment—but there's more. It's also vital for our understanding of prevention of the intergenerational transmission of ACEs. It opens avenues for change that can echo through future generations.

In summary, the journey towards addressing ACEs and breaking their cycle through generations is not merely an academic exploration but a vital, attainable mission. By leveraging the insights into neuroplasticity, we are no longer confined to static models of the human brain. Instead, we're presented with an organ of immense adaptability, capable of rewriting the neural scripts that govern behaviour and well-being throughout one's lifetime. This transformative quality of the brain becomes particularly crucial when contending with the persistent, often debilitating, effects of ACEs and stress.

However, the landscape of this transformation is ever-evolving, especially in the context of contemporary challenges like parenting in the digital age. As technology deeply embeds itself in our lives, it becomes another variable in the complex equation of human development and neuroplasticity. Parents and caregivers are thus tasked with guiding the next generation through both the potentials and pitfalls of the digital world, making informed choices that align with positive neural development. This confluence of technology and neuroscience adds another layer of responsibility for our society. We must ensure that the digital environment is used for healing, rather than as an exacerbator of ACEs.

Let Your Silence and Nature Do Some of the Heavy Lifting

Let's take a deep breath and pause for a moment of silence. In the silence, find some relief from the constant push and pull of life and the overwhelming nature of parenting overload and guilt. In this pause, we allow ourselves to find an easier way through the quagmire of parenting advice in the digital age. This is a call to be silent and access simple parenting tools first—tools that often disappear under the stress and pressure of daily routines.

John Bray, a school principal, underwent a transformative shift while working in Aboriginal communities. The community Elder invited him to a "yarning circle", a traditional gathering where stories, wisdom, and ideas are shared. As John sat down, he noticed something different. There were long pauses, periods of profound silence. At first, the silence made him uncomfortable; he was eager to fill it with words, suggestions, and questions. But the Elder gently guided him. "In our culture, we listen to the silence. It's where the land speaks to us, where wisdom hides."

Intrigued, John decided to follow suit. As he sat in silence, he became acutely aware of the world around him—the rustle of the leaves, the distant laughter of children playing, and even the nuanced expressions on people's faces. But, most importantly, he began to listen in a way he never had before.

Over time, John found that the silence enabled him to absorb the community's wisdom at a deeper level. He began to understand the complex relationship the community had with the land, and the systemic challenges they faced. He also realised that many of the "solutions" he had initially considered to be helpful were either irrelevant or could have been disruptive to the community's way of life.

When he returned to his work in urban settings, John carried this lesson with him. He began implementing moments of silence in staff meetings and family discussions, encouraging people to listen before they spoke. It was challenging at first; the silence often met with discomfort. But gradually, people began to see its value. In silence, they found clarity, deeper understanding, and, often, unspoken solutions to problems that words had failed to address.

John Bray credited the Aboriginal communities for teaching him the art of listening through silence—an invaluable lesson that he has taught me, and which is hard to do.

The power of silence is frequently underestimated, especially when it comes to parenting in challenging situations. Silence, when used strategically, can serve as a tool for easing tension and fostering a conversation.

Chapter 12: RE-PARENTING

Parents can significantly influence the emotional climate of a conversation, particularly during heated moments. Try sitting in a circle with your family and taking a deep breath, which not only helps to calm your own nerves but also serves as a subtle signal to your child that the space is safe for a chat. Putting away distractions like phones and laptops is crucial in these moments. The act of setting aside a device demonstrates that your full attention is on the conversation, reinforcing its importance and showing respect for your child's feelings. This simple action can make a significant difference in how effectively you can defuse tension.

Eye contact, often underestimated, plays a pivotal role. Making gentle eye contact can help to establish a connection and indicate that you're actively listening, without the need for words. Lastly, the tone of your voice and the way you carry yourself physically can also serve as powerful tools for setting the emotional tone. A soft yet clear voice, devoid of harshness or criticism, can convey openness. Adjusting your body posture to lean in slightly can also send a message of attentiveness and care. Try to repeat back what you heard, to ensure understanding. By mastering these non-verbal cues—deep breathing, attentive posture, elimination of distractions, thoughtful eye contact, and conscious voice modulation—you're not just managing the immediate situation. You're also teaching your child important emotional regulation skills and setting the stage for more open, respectful communication in the future.

When all else fails, go for a drive together, or sit alongside your children, and as much as possible, let them speak. John went on to describe that even the most challenging children who have been through unspeakable traumas in their life will open up in the space we create for them. Where we remain silent and attentive.

This is where the magic of BEING SEEN is felt. There is nothing like the feeling of being there when someone feels safe and calm and speaks up for the first time. Parents, you have the power; you are in charge. Children are children and cannot make good decisions when they are being groomed online or asked to do things that are inappropriate. We forget simple actions when we are under stress. It is because when we're stressed, our fight-or-flight response kicks in, narrowing our focus to the challenges at hand, and pushing aside anything that doesn't seem directly related to solving the problem. This survival mechanism, while useful in certain situations, doesn't serve us well when the stressors are not life-threatening but are rather the everyday strains of work, parenting, and managing a household. Further, each of us has a different speed of reactions based on our early life experiences and how we were parented. In fact, it is in these stressful moments, that you become fully aware of how you were parented.

When you're at your wits' end, overwhelmed by children who seem more responsive to screens than to you, reclaiming parental authority can feel like an uphill battle. As they push boundaries and exhibit challenging behaviours, the urge for a quick and effective solution intensifies. Ironically, it's during these high-stress moments that we often overlook simple remedies. Our judgement gets clouded by a whirlwind of responsibilities, looming deadlines, and the noise of digital distractions affecting both us and our stressed children.

Simple acts like stepping outside for fresh air, or taking the dog for a walk or the children to the park can serve as powerful antidotes to stressful and charged moments. These practices require little to no effort and are almost always accessible, yet they often fall by the wayside when we're overwhelmed. It's as if the rising tide of stress blinds us to the very lifelines that could help us navigate through it. Remembering to return to the basics during stressful times is sometimes the best and only thing you can do when everything feels like too much. Turn off the screens, step outside, breathe deeply, and reconnect with the elemental joys that life—and nature—have to offer. It's often these simple things, easily forgotten yet profoundly impactful, that recalibrate us and provide the resilience needed to face life's complexities.

Chapter 12: RE-PARENTING

CASE STUDY – Anna and Mario

A crumpled to-do list and the clock ticking closer to bedtime, Anna and Mario felt a wave of despair wash over them. It was as if they were stuck in a never-ending loop of stress and disappointment, both in themselves and in their children.

But then, they remembered to let nature do the heavy lifting in resetting the brain's stress response system. With deep breaths, they decided. *Enough is enough.*

"Kids, put on your shoes. We're going for a walk," Anna announced, disconnecting the iPads from their hands. Ben and Diana looked up, confused but intrigued.

"Why?" Ben questioned, his voice tinged with the usual pre-teen scepticism.

"Because we all need a break and some fresh air," Mario replied.

Shock and hesitation were in their eyes, the pull of digital screens still strong. But Anna and Mario stood their ground. The initial few steps outside were filled with resistance and subtle whining. As they walked down the suburban streets and finally entered the park, something magical happened. The further they got from their home—and from their digital distractions—the more they seemed to come alive. The twins started noticing the butterflies, the fallen leaves, and even started competing to find the most unusual rock or stick. As they walked, Anna and Mario felt their own stress levels drop. The weight of their responsibilities seemed to lessen, for the first time in a long while. The digital world, with its incessant notifications and soul-draining screens, felt far away.

"Look, Dad, a puppy!" Diana shouted.

Mario smiled, his heart lifting. "It's so fluffy, isn't it?"

By the time they returned home, the atmosphere had changed entirely. Dinner was still unfinished, and homework was still pending, but none of that seemed as insurmountable as before. And, more importantly, they felt like they had reclaimed something invaluable: their own well-being and connection with their children. From that day on, walks became a non-negotiable part of their daily routine. The iPads still existed, of course, but they no longer held the same hypnotic power over Ben and Diana. They had found a lifeline, a simple yet powerful antidote to the chaos. And all it took was a step outside.

So the next time you find yourself overwhelmed and at your wits' end, remember Anna and Mario's story. Sometimes the simplest solutions are the most effective; we need some simple solutions when we feel like everything is outside our control. Remember the last time you were in nature—perhaps it was a walk through a forest or a day spent by the ocean. The sounds of birds chirping, waves crashing, or leaves rustling in the wind likely brought an inexplicable sense of calm and grounding.

The inherent tranquillity of natural settings often lowers stress levels and softens emotional defences, paving the way for deeper, more meaningful interactions. So, the next time you find yourself at an impasse, consider switching off the devices and taking a family excursion into nature, to the park, in a garden—it might just be the perfect antidote to the complexities of parenting in the digital age. Amidst the parenting rollercoaster, isn't it a breath of fresh air to realise there are some simple, timeless tools right at our disposal? The calming embrace of nature has always been here, offering us a steady hand in our parenting journey. I can almost hear a collective "phew" from all of you. *Finally, a straightforward tip!*

In our next chapter, we'll explore the how to navigate the technology world with more ease and less panic. Just as each child is unique, every parent has their own way—and embracing yours might just be the secret sauce to a joyful parenting experience.

Curious to know more? Let's dive in together, with renewed energy and excitement.

Time for Reflection

This exercise aims to help you understand and harness the power of neuroplasticity to positively influence your mental well-being and overcome the echoes of past experiences. Consider sharing your goals with a trusted friend or family member who can encourage your journey. Remember, you don't have to do this alone. Seeking professional guidance or counselling can provide support if it is needed.

Understanding Your Past

- Reflect on significant events from your past that you feel might have impacted your current behaviour, emotions, or thought patterns. Write them down.
- Consider how these events might be influencing your reactions to current situations or challenges.

Awareness of Current Behaviour

- Observe and note down behaviours, thoughts, or emotional reactions that you believe are linked to past events.
- Are these patterns helpful, or do they hinder you in some way?

Envisioning Change

- Based on your understanding of neuroplasticity, remember that your brain can adapt and change.
- What are some positive behaviours or thought patterns you'd like to cultivate? Write them down.

Celebrate Progress

- Every week, reflect on the changes you've observed.
- Celebrate small victories and remember that change is a process.

Chapter 13

UNLOCK TECH-SAVVY SUCCESS

Family Tech Plan, Tech-Free Zones, Social Media, and Screen Time

In Brief

Children born after 1996 are the first generation of kids to have cell phones and social media throughout school. As we see depression, anxiety, and suicide rates skyrocket alongside mobile phone and social media usage, we can see that there is something new happening. Unlike the introduction of newspapers or television, the arrival of technology is making change happen at an exponential rate—so quickly that our brains can't keep up.

Technology like social media feeds into our need to belong. But our brains haven't evolved at a rate that coincides with the rate of developments in this area. How can we feel belonging when we're comparing ourselves to billions of other people? How can we feel safe when we're constantly reminded about how unsafe the world can be? Because the digital age presents new challenges for parents that demand solutions that go beyond traditional conversations about "the birds and the bees" or setting curfews. Parents are engaging in a virtual game of "Whac-A-Mole" across various platforms, as concerned parents try to help each other navigate the overwhelming landscape of social media and the plethora of devices at their children's fingertips.

Every parent I encounter seems to be grappling with the challenges of technology, from screen time to the exposure of inappropriate content. In our current digital age, managing smart devices has become a primary parenting concern, overshadowing issues that may have been central for previous generations. The screen has essentially become the latest arena for child-rearing challenges.

The genie is out of the bottle, and there's no turning back. We must confront the stark reality that smartphones and social media platforms pose significant risks to young, developing minds. For instance, scientific research shows that the prefrontal cortex, the brain region responsible for decision-making and impulse control, doesn't fully mature until the age of 25 in males. This makes the allure of smartphones, games, and other digital distractions particularly hazardous.

Being Tech Savvy Doesn't Equate to Possessing Wisdom Or Street Smarts

Given that today's youth are often dubbed "digital natives", fully immersed in this technology-savvy world, the question becomes: how do we navigate parenting in the digital age? It's essential to recognise that being technology savvy doesn't equate to possessing wisdom or street smarts. Parents, although they may be "digital immigrants", bring a wealth of experience and nuanced understanding of human behaviour that young people simply haven't had the time to develop.

This is why as a collective we need to build a next-level strategy—one that combines life wisdom and past parenting styles with a tech plan, digital literacy, and tech-free zones, to guide children effectively in the digital age. But here's the game-changing truth: *you have an unparalleled edge—you're not just a parent; you possess an arsenal of wisdom, life experiences, and community support that can* effectively counter the digital pressures exerted by both technology-savvy children and profit-driven technology companies.

CASE STUDY – Isabella

Isabella found herself sitting at the dinner table, a sumptuous meal spread out before her and her family. Her husband was scrolling through emails on his phone, her teenage son was lost in a multiplayer game on his tablet, and her younger daughter was engrossed in a virtual art class on her laptop. The food grew cold as each family member interacted with their devices; their virtual worlds far more captivating than the reality of the family dinner before them.

Isabella sighed as she looked around the table. It was as if a digital wall had been erected between each member of her family, separating them despite their physical closeness. Ironically, each was more connected than ever before—just not to each other. Her husband was plugged into his global network of colleagues, her son was battling it out with friends from school and strangers from across the world, and her daughter was learning from a renowned artist based in another country. Yet, in that moment, Isabella couldn't shake off the feeling that something essential was missing.

She missed the days when dinner was a time for sharing stories from the day, for laughter and bonding, a sacred family ritual that technology hadn't yet invaded. She yearned for the moments when "connectivity" meant eye contact, a touch, a hug, or a meaningful conversation. As she gazed at her family, each absorbed in their own digital universe, Isabella realised that while technology had succeeded in connecting them to a broader virtual world, it had also subtly disconnected them from their immediate reality—their family.

It was then that Isabella knew something had to change. The challenge was clear: how to harness the benefits of this incredible digital age without letting it erode the very real, human connections that make life truly meaningful? It was a challenge that she knew many families were facing, a modern paradox of being more connected and yet feeling less connected than ever before.

Determined to reconnect her family, Isabella took a deep breath and made a tech strategy for them. Isabella had done her homework. She started with digital literacy. She would create a weekly family meeting to discuss the apps and platforms they were using—not to snoop on each other, but to understand and help one another navigate the digital world safely. To tackle the more sensitive issues, Isabella initiated age-appropriate conversations about online safety and etiquette and the potential risks, like encountering inappropriate content. Knowledge is power, and we should all be aware of both the good and the bad that can come from our digital lives.

Next, Isabella started learning how to apply the concept of Socratic parenting. She wrote down and started learning how to pose open-ended questions like "What was the most interesting thing you saw online this week?" or "How do you feel when you spend a lot of time on your devices?" These questions served as conversation starters, creating a safe space for everyone to share their digital experiences and concerns. Isabella also designated tech-free zones in the house—the dining room, bathrooms and the bedrooms—and established tech-free times, particularly during meals and an hour before bedtime. "These will be connection times, a time for activities that bring us closer, like sharing meals, playing board games, or simply talking or reading together."

Isabella made it a point to consistently apply these strategies. Over time, the family noticed a change. The tech-free zones and times became moments of connection, the weekly digital discussions became something to look forward to, and the open questions not only kept the conversations flowing but also deepened their understanding of one another. A few months later, Isabella found her family around the dinner table again. But this time, the atmosphere was different. The food was warm, the conversations were engaging, and the devices were nowhere in sight. Her family was not just physically present but emotionally connected as well.

As they shared stories, laughed, and looked each other in the eyes, Isabella felt a sense of accomplishment wash over her. She realised that while the digital age had presented a complex challenge, it was not insurmountable. By proactively adapting and implementing focused strategies, she had successfully navigated the digital age. She had ensured that her family felt seen and heard—not just by their device cameras, but by each other.

And so, Isabella's family proved that even in a world buzzing with digital distractions, meaningful connections are possible. It takes effort, open dialogue, and a commitment to adapt, but the reward—a family that's truly connected—is more than worth it.

While issues like education, socialisation, mental health, nutrition, sleep and exercise remain critically important, their impact is significantly influenced, if not shaped, by the digital environment in which our children are immersed. That's why this chapter begins by tackling the omnipresent screen. Whether it's the addictive pull of social media, the potentially harmful content lurking online, or the way screen time can displace face-to-face interaction, the influence of technology is so pervasive that it can negate efforts in other important areas of parenting.

If children are ensnared by what's happening behind the screen, breaking free becomes the first step in addressing a host of other issues. Therefore, equipping parents with the tools and strategies to manage this digital landscape effectively is not just a starting point—it's a cornerstone for all the parenting challenges that follow. Remember, the goal is for your children to know you are a confident parent who has mastered the skills to listen to them. When we ourselves are stressed and under pressure we lean on what our parents taught us in how to respond to children, but this is not always the best approach. Instead, we need to have a set of tested tools and an arsenal of digital literacy up our sleeve that can be relied on in the heated and distressing moments.

We'll delve into actionable strategies like finding supportive parental social networks, crafting family tech plans, and bootcamping digital literacy to fortify your tech knowledge and understand how your children are using technology rather than imagining they are seeing what we are seeing. By leveraging these tools, you're not just setting boundaries; you're cultivating an environment of open dialogue, mutual respect, and shared responsibility.

Some Ideas to Get Started

This is a hard one, but we need to start the conversation about giving unfettered access to smartphones to your children until they reach the age of 13. At a minimum, adjust the privacy and security settings on computers, smartphones, and tablets so that you can monitor usage and conversations online.

CASE STUDY – Carolyn

Carolyn had had one of those days where everything seemed to go wrong. Her children, Phillip and Nicole, were more engrossed in their iPads than in anything she had to say. Dinner was a battle, homework was a disaster, and she felt like she was losing control. As she stood in the kitchen, staring at her children who were, yet again, glued to their screens, she felt her stress levels skyrocket.

Carolyn walked over to the living room, unplugged the wi-fi router, and waited. It took about 30 seconds for Phillip and Nicole to look up, bewildered.

"Family meeting, *now*," she declared.

The kids groaned but complied.

"Alright," Carolyn began, "I've noticed we're all a bit on edge and glued to our screens. So here's the deal. For the next hour, no screens. We're going to play a board game. Whoever wins gets to pick dessert for tonight."

Phillip and Nicole looked at each other, then at their mum. It was a simple solution, almost too simple. But they were intrigued.

That hour turned out to be the most peaceful Carolyn had felt in weeks. They laughed, and, for those moments, the digital world was forgotten.

The pervasive influence of digital devices in our lives is undeniable. Yet the narrative of Carolyn and her children underscores the power we possess to reclaim real-life moments of connection, even amidst the digital tide. As parents, fostering healthier brain development isn't just about understanding neuroscience or setting strict technology rules. It's about recognising the moments when we need to unplug, redirect, and immerse ourselves in the tactile, real world with our loved ones. The journey to promote better brain health for our children doesn't have to be complex. Sometimes, all it takes is turning off the screens and turning towards each other. As we move forward, let's remember the simplicity of board games, conversations, and shared laughter—for these are the timeless ingredients of connection, growth, and well-being.

We might believe that we lack the influence to bring about significant change to keep our children safe online in the digital age. Yet, seeing the impact of the Matildas and other teams during the FIFA Women's World Cup Australia & New Zealand 2023™, and how it elevated global appreciation for women's football, made me realise that every one of us holds the potential to spark significant societal shifts. As I sat in the stadium, I felt the wave and witnessed firsthand the energy from the field resonating with the minds of those watching. Fans and spectators become part of this dance, their emotions and connections mirroring those on the field. For a couple of hours inside that stadium, I glimpsed what a world interconnected by a shared purpose—a genuine desire for others to succeed—might feel like. It was a fleeting taste of unity and human connection.

The memory lingers, a reminder of the power that unified human connection holds for transformative social change. It's imperative that we leverage the strength of our social networks to spearhead a movement against technology addiction, much like the successful campaigns against smoking. While the digital age presents us with unprecedented challenges, it's crucial to remember that humanity has historically demonstrated resilience and adaptability in the face of transformative change. Just as we've navigated past societal shifts, we will, in time, develop effective strategies and norms to handle technology in a manner that serves our best interests.

How to Create Healthy Boundaries Around Technology

As you lie in bed doom-scrolling, 30 minutes may pass, and you end up missing your first meeting of the day. Or, after looking at everyone else's holiday celebrations, the FOMO sets in, and you no longer feel happy or motivated. By becoming more aware of what you pay attention to first thing in the morning and throughout your day, you're making a choice regarding your brain health. If we constantly feed our brain input from social media sources, we're going to feel overwhelmed and anxious, and as if we're never doing enough or that we *are* never enough. The truth is, we can't feel like we belong, and we can't feel safe. In many ways, technology works against our innate needs as tribal creatures under the guise that it helps foster connection (which, in some cases, it absolutely does!). But for the developing brain and improved mental health, we need in-person interactions to learn and develop in a healthy and positive way. We need touch and connection without distraction.

If we look at the statistics, we see a strong correlation between smartphones becoming our everyday tools and mental health diagnoses rising. The idea that having social media at arm's reach and declining mental health are linked isn't new, by any means! The research also shows that girls are particularly susceptible to the negative impacts of social media. The statistics further show an increasing number of teens meeting up in person with their friends only once a month or less due to the rise in digital devices. Yet face-to-face interactions, as we know, are important.

While many experts rack their brains regarding how we can combat these problems, we can create a few simple rules for our own families and children, such as:

- No phones in the bedroom and no mobile phone use an hour before bed and an hour after awakening.
- No phones or digital devices at the dinner, lunch, or breakfast table.
- No mobile phone use before school, during school, or when completing homework.
- No digital devices during family time, such as movie night, hiking, or other activities.

Know the dangers of social media. Set limits. The impacts of social media usage are wide and profound, not just for our children but also for us, as parents and adults. As parents, we are the models for our children regarding healthy digital device usage. This means we should display the behaviour we expect from our children. It also goes without saying that having a discussion regarding safety online is imperative (and ensuring you approve all app downloads may also be beneficial until your child reaches a certain age). Focus on prioritising in-person connections with your children and creating activities together that nurture the serve and return relationship. This way, you'll guide your children towards healthy and happy futures where they learn safe and healthy boundaries regarding technology and social media.

Use Your Online Parental Social Network

Remember, the digital world is merely a new context for age-old parenting challenges. And just as parents have done for generations, you too can adapt, learn, and thrive. So, let's harness the collective power of parental wisdom to navigate this new digital frontier. As parents, you've navigated complexities that extend far beyond the digital era: interpersonal relationships, ethical dilemmas, emotional intelligence, and the long-term consequences of actions. Street smarts—acquired from real-world experiences rather than from a

YouTube tutorial—involve understanding motives, recognising risks, and making judgements that consider the broader context. Parents have another invaluable resource: their social network. Unlike children, who may rely solely on peers for advice, parents have access to a diverse and experienced network that includes other parents, educators, and professionals. Leveraging this collective wisdom can provide insights and strategies for tackling the challenges of parenting in the digital age.

Wisdom circles that leverage the parental social network can be invaluable forums for shared learning and support. These circles, either formed among friends or joined online, serve as collective think tanks where parents can exchange insights, research findings, and effective strategies for tackling challenges like screen time, online safety, and digital etiquette. There are many online Facebook groups such as "Parenting in a Tech World" and many others. Having a support network is critical for sharing ideas and helping each other.

For example, a parent concerned about her teenager's escalating screen time brought up the issue in one such group. Another parent in the group shared a creative "Tech-Free Tuesday" initiative they had implemented in their household, where the entire family unplugged for the evening to engage in board games and face-to-face conversations. Intrigued, the concerned parent adapted this strategy for her family, and found it remarkably effective in reducing screen time and enhancing family bonding. The collective wisdom of the group had offered her a solution she might not have considered on her own. By pooling their individual experiences and insights, parents are often better equipped to navigate the challenges of the digital world.

Exploring Common Digital Parenting Challenges: A Community Perspective

As technology surges ahead, many parents find themselves lagging, while their children adapt with astonishing agility. This digital gap raises a multitude of concerns for parents, from online safety and screen time to potential exposure to inappropriate content. Through sharing experiences and solutions, parents can feel less isolated and more equipped to guide their children through the complexities of the digital landscape. Let's delve into some common questions raised by parents in this digital age and explore the collective wisdom and strategies employed by the community in response.

Question. Despite setting downtime on my child's phone for 9 pm, she can still access Snapchat. What am I missing?

Community Insight. Many parents have encountered this glitch, often due to app updates. A popular solution is to adjust the specific app limits for Snapchat. Alternatively, some parents switch off their home Wi-Fi at 9 pm to ensure all online activity ceases.

Question. I've been open with my 10-year-old about internet dangers, but he was still found searching for inappropriate content. He's now blaming his older brother, and while we've assured him there's no punishment, I'm worried about his deceit as much as the searches. We already monitor tech use. What should we do?

Community Insight. This is a concern for many. Some parents recommend involving the child in establishing internet safety rules, which encourages responsibility and openness. Also, many schools have resources for safe tech use; partnering with them can reinforce the message at home.

We need to activate next-level strategies to effectively navigate the complex digital landscape with our children—strategies that use our age-old wisdom circles. In facing these digital challenges head-on, together we can lean into collective wisdom and craft a roadmap for the next generation.

Bootcamp Your Digital Literacy

Just as venturing into a new country with a different language necessitates learning key phrases like *bonjour* in French, *hola* in Spanish, *ciao* in Italian, and *konnichiwa* in Japanese, parenting in the digital age demands that we acquire a new set of skills and "digital phrases" if we are to effectively communicate and connect with our children. We can learn from the children and start to speak their language. We need to be using the same platforms, games and technologies as they do, so we understand how they work and what is happening online.

Creating quality time for regular open conversations allows families to learn what platforms the children know about and are using and ensure that parental understanding remains relevant and practical. Creating moments of learning through regular family discussions on topics like digital use, online behaviour, and digital ethics is not just a rule-setting exercise; it's a fertile ground for mutual learning. For example, during one such family conversation, a mother shared her concerns about the hidden dangers of oversharing personal information online, drawing parallels to her own experiences of trust and caution in the real world of being scammed. Her teenage daughter listened intently and reciprocated by showing her mum how to tweak privacy settings on social media platforms for better security. This became a teaching moment for both. The mother learned something practical about digital safety, while her daughter gained a deeper understanding of why caution is important. Such conversations serve as a two-way street for transferring knowledge, allowing parents to impart their life wisdom even as they adapt to the digital realities their children navigate with ease.

Here's why. When children feel they can discuss their ideas and feelings without immediate judgement, they are more likely to share their online experiences with parents. This openness can provide parents with valuable insights into potential issues, such as cyberbullying or inappropriate content, before they escalate. The non-judgemental atmosphere cultivates a sense of trust. If children encounter something unsettling or dangerous online, they'll be more likely to approach their parents for guidance, because they expect an open, understanding response. Guiding children to discover solutions for themselves equips them with the skills to make good decisions independently. This ability is crucial online, where parents can't always be present to guide every click or interaction.

Critical thinking skills help children assess the risks and rewards of different actions in a nuanced way. Whether they are deciding to share personal information or download an app, a child equipped with good problem-solving skills is better positioned to make safer choices. Children will eventually grow up and manage their digital lives without parental oversight. The earlier they learn to think critically and make good decisions, the better prepared they will be to navigate the digital world safely. The digital landscape is continually changing, presenting new challenges and scenarios. A child who has learned to think critically can adapt more easily to new situations, making them less vulnerable to online risks.

Have a Family Tech Plan

A Family Tech Plan is a structured approach to managing technology within the household, aiming to balance screen time, ensure online safety, and promote healthy digital habits among family members. This plan serves as an agreement between all members of the family, outlining the rules and guidelines for using technology. The benefits of having a shared tech plan are that it helps children and adults become less stressed and minimises the risks associated with excessive screen time and online dangers. Most importantly, it encourages more quality time spent together as a family, free from digital distractions. This serves as the foundation for lifelong learning, mental health and well-being. Creating and sticking to a Family Tech Plan can help families navigate the challenges of the digital age while reinforcing positive behaviours and habits around technology.

Key Components of a Family Tech Plan

Screen-Time Limits. Establish daily or weekly limits for non-educational screen time for children and adults alike. Specify times when screens are off-limits, such as during meals or an hour before bedtime.

Devices and Tech-Free Zones and Times. Designate certain areas in the home, like the dining room, bathrooms and bedrooms, where electronic devices are not allowed.

Content Guidelines. Set rules on the types of content that can be accessed. Use parental controls to block inappropriate material and monitor usage. Make sure you can access the content at any time on a child's device.

Online Safety. Teach and implement basic online safety practices such as not sharing personal information, recognising phishing scams, grooming and sextortion and the importance of strong passwords.

Educational Use. Define what constitutes educational screen time, like research for homework or educational games, and how it differs from recreational screen time.

Social Media Rules. For older kids who use social media, establish guidelines about what is acceptable to post, and discuss the potential ramifications of sharing personal information online.

Family Tech Time. Allocate time for family activities that include tech, such as a family movie night or playing an educational game together. This encourages positive and communal screen time.

Emergency Use. Clarify the rules for using smartphones and devices for emergency situations. Make sure every family member knows how to contact each other and where to find important numbers.

Check-ins and Adjustments. Regularly review the tech plan to see if it's working or if any rules need to be adjusted. This could be done during a monthly family meeting.

Consequences. Clearly outline the consequences for breaking the rules, whether it's loss of screen time or confiscation of devices.

Real-Life Example – *The Johnson Family*

The Johnson family found themselves struggling with tech boundaries. Their 14-year-old son, Tim, was spending more hours than they were comfortable with playing online games, while their 12-year-old daughter, Erin, was becoming increasingly engrossed in social media. The Johnsons decided it was time to draft their "Family Tech Plan". This wasn't just a list of rules; it was a negotiated document that included input from Tim and Erin, making them more invested in adhering to it. The plan was placed on the family bulletin board and revisited quarterly for updates. All members agreed and were part of the plan.

But the Johnsons took it a step further with "Transparent Monitoring". They installed a parental control app, like the Bark app, but sat down with Tim and Erin to discuss why it was necessary and what it would monitor. Every Sunday, they would look over the week's online activity together. When Tim visited a website that had raised a red flag, the family talked it through. Tim explained he was researching for a school project, which led to a broader discussion on reliable sources. The experience was eye-opening for everyone and turned into an educational opportunity rather than a punitive moment.

Create Tech-Free Zones and Times

Tech-free zones can serve as sanctuaries within the home, promoting face-to-face interaction and social connection. Designating areas like the dining room, bathrooms and bedrooms as tech-free zones can transform these spaces into havens for quality family time. Specifying tech-free days and times can have the same effect.

We are used to trying to reduce alcohol using Febfast, Dry July and Octsober. Implementing regular tech-free periods like "Tech-Free Tuesday" or a weekend "Digital Detox" can be a powerful way to shift the focus back to family connections. For example, the Williams family initiated a "Tech-Free Tuesday" after noticing their evenings were consumed by individual screen time. At first, their teenage daughter was sceptical, lamenting the loss of her social media scroll time. However, as they spent their first tech-free evening after cooking dinner together, the atmosphere in the house transformed. Laughter filled the rooms, and everyone seemed genuinely engaged with each other. The unexpected benefit? Their daughter admitted she felt less stressed and more "present".

To keep the momentum going, the Williams family established bi-weekly "Tech Talks". These are open forums where they discuss new apps, online challenges, and even dissect news stories about technology's impact on mental health. They've created a safe space where everyone, especially the kids, can openly share their experiences and concerns without fear of judgement. This ongoing dialogue has not only kept the parents updated on tech trends but has also empowered the children to think critically about their digital interactions.

Arm Yourself with Open Questions and Practise Them

For parents, it's all too familiar: those moments when emotions run high and our child's behaviour feels overwhelming. In these times, it's crucial to be equipped with the right approaches to help navigate the situation smoothly. Equip yourself with a few open-ended questions, which can be invaluable in such moments. Below, you'll find examples suited for various situations. These questions aim to prompt more thoughtful responses than simple "yes" or "no" answers, encouraging children to reflect, share their emotions, and work together towards solutions. If you are not used to this method, try learning some open-ended questions and practising one or two in less heated moments. Have up your sleeve a set of questions that can't be answered with a simple "yes" or "no". For example, try "How was your day?" instead of "Did you have a good day?"

Here are some open-ended questions to try for addressing the most important and challenging issue facing parents right now.

Understanding Screen-Time Choices

"Can you help me understand what draws you to this particular game/show/app?"

"What do you like most about spending time on your device?"

"How do you feel when you're not able to use your screen?"

Discussing the Rules

"What are your thoughts on our current screen-time rules?"

"Why do you think we have rules about screen time?"

"How do you feel about the amount of time you spend on screens compared to other activities?"

Addressing Rule-Breaking

"I noticed you've been using your device more than we agreed. Let's talk about that?"

"What led you to go around the rules we set?"

"Please tell me what you're trying to accomplish or experience that you feel you can't do within our current guidelines?"

Exploring Emotional Responses

"How do you feel when you have to turn off your device?"

"How do you think your mood changes with the amount of screen time you get?"

"What emotions do you experience when you think about reducing screen time?"

Focusing on Solutions and Alternatives

"What activities could you see yourself doing if you spent less time on screens?"

"How can we make our screen-time rules more effective?"

"What would a reasonable screen-time schedule look like for you?"

Enhancing Mutual Understanding

"What is there about screen time that you wish I understood better?"

"What could we do as a family to find a balance between screen time and other activities?"

"How do you think we could make screen time a more positive experience for everyone?"

Dealing with the Risk of Explicit Content and Grooming

Remember you are not alone, and most parents and educators feel lost, are in shock, and have no idea what to do. Please take a deep breath and approach the following statement with an open heart and compassionate understanding. This isn't something any parent wants to hear, nor is it something many are told: *smartphones, social media, and games are not safe spaces for young children.* The situation has escalated far beyond what many of us anticipated.

Many of us operate under the assumption that our children use these technologies in the same innocent way we do. But the reality is starkly different. Speaking from personal experience as a parent, I wish I'd known this information when my own children were teenagers. Disturbingly, as far back as 2015, teenagers were sharing explicit images with each other online and resorting to blackmail. Today, we are facing an alarming increase in age-inappropriate material affecting young children.

Make no mistake: just because something is available online does not make it safe. Research indicates that exposure to inappropriate or explicit content at a young age can have long-lasting effects on a child's brain development and emotional well-being. So, let's embrace this challenge with urgency and determination. We owe it to our children to ensure their physical and mental health remain intact. Using open-ended questions to keep respectful, non-judgemental lines of communication open with our children can help us to help them. Let's equip ourselves with the knowledge and tools needed to steer them away from harm's way.

We didn't see this coming; we didn't expect our children to be exposed to adult content at such tender ages. Yet here we are, and the first step in navigating this challenging terrain is acknowledging the problem. Let's do this together, because I know parents love their children and want the best for them. It's a hard pill to swallow, but our awareness and action can make all the difference in safeguarding our children's futures.

Identifying Signs and Open Communication

The first step in combating grooming is education—both for parents and children. Parents should be aware of the signs of grooming, which can include a child becoming secretive about their online activities or showing a sudden emotional attachment to an online friend.

Here is a list of some main signs that may suggest a child is being groomed online:

1. **Excessive Secrecy.** The child becomes overly secretive about their online interactions, often hiding screens or quickly switching tabs when someone comes near.

2. **Unexpected Gifts.** The child receives gifts or packages from someone they've met online, which can be a tactic used by groomers to gain the child's trust.

3. **Changes in Behaviour.** There may be notable shifts in mood or behaviour, such as becoming withdrawn, upset, or displaying outbursts after using the internet.

4. **New Electronics or Accounts.** The child suddenly has new electronic devices or online accounts that they didn't have before, especially if there is no clear explanation of how they got them.

5. **Increased Online Activity.** There is a significant increase in the amount of time the child spends online, especially if it is in private or during unusual hours.

6. **Adult Friends.** The child has new "friends" online who are significantly older, or there is evidence of conversations with adults you don't know.

7. **Unwillingness to Discuss Online Friends.** The child is reluctant or defensive when asked about their online friends or activities.

8. **Concerning Content.** You discover sexually explicit or otherwise inappropriate content on the child's device.

9. **Withdrawal from Regular Life.** The child may become increasingly disengaged from their regular friends, family, or activities they used to enjoy.

10. **Phone Calls from Strangers.** The child receives phone calls from numbers you don't recognise or makes calls to numbers you don't know.

11. **Use of Sexual Language.** The child uses sexual language that you wouldn't expect them to know, indicating they may be learning it from someone online.

12. **Emotional Manipulation.** The child might show signs of emotional manipulation, expressing feelings of guilt or shame that someone online has instilled in them.

It's important to note that these signs can sometimes be indicative of other issues, but if a child is showing several of these behaviours, it could be a reason to look closer and have an open and supportive conversation with them about their online interactions.

Open communication channels between parents and children are crucial. Regular, non-intrusive discussions about online interactions can go a long way in early detection and prevention. In the 2020s, digital literacy goes beyond understanding how to use a smartphone or browse the internet; it includes understanding the nuances of online behaviour and the potential risks involved. Teach your children how to safeguard their information online and how to be critical of online interactions when an individual starts asking for too much personal information.

Real-Life Example – *Sarah and Grooming Red Flags*

Sarah, a mother of a 13-year-old, noticed that her daughter was spending a lot of time chatting online and seemed overly invested emotionally. After an open and frank discussion, she learned that her daughter was indeed in contact with someone whose behaviour raised red flags. Immediate steps were taken to cut off contact, and Sarah used the experience as a teaching moment not only for her daughter but also for her community, raising awareness about the dangers of grooming.

In today's rapidly evolving digital landscape, parents face unprecedented challenges in discussing sensitive topics like technology use and sexuality with their children. The "smoke detectors" in our brains are often on high alert due to the stressors of modern life. Therefore, learning to manage our stress response is crucial for effective parenting, especially when discussing complex topics like technology and sexuality. As we navigate through the complexities of parenting in the 2020s, it is essential to update our skill set to include digital literacy and psychological understanding, so we can protect our children from new types of risks like grooming. With open communication, education, and the responsible use of technology, we can provide a safer environment for our children to grow and thrive.

This section aims to equip parents with effective strategies for approaching these delicate conversations, backed by neuroscience and real-life examples from parents who have successfully navigated these waters.

Real-Life Example – *Emily and Screen Time*

Emily, a mother of two teenagers, found herself in a constant battle over screen time. Her stress levels were through the roof, clouding her ability to have a rational discussion with her kids about responsible tech use. Every conversation turned into a shouting match, leaving both parties frustrated.

Emily decided to take a step back and reflect. She started practising techniques to calm her smoke detector. The next time the screen-time issue came up, she was able to discuss guidelines and consequences calmly, and her children were more receptive.

Talking about Technology and Sex in an Age-Appropriate Manner

We hold the power to protect our children from the dark corners of the digital world. As parents, we must educate ourselves on how technology and too-early viewing of sexual content impact brain development and young minds. Only then can we take charge, set boundaries, and create a safe digital environment for our children to thrive.

The struggles of discussing technology and sex with children are very real, and intensified by our natural stress responses. However, we can develop the skills needed for these crucial conversations. In doing so, we empower our children and also foster a family environment where difficult topics can be addressed openly and constructively.

Starting early and maintaining an ongoing dialogue about technology is crucial; it should never be a "one-and-done" conversation. Tailor your discussions to be honest yet age appropriate, ensuring that the information resonates with your child's level of understanding. Additionally, it's vital to cover the topics of consent and boundaries, which are often neglected but are key components of sexual education. By continually revisiting these subjects, you can adapt your guidance as your child matures and encounters new challenges in the digital world. Here are real-life examples and constructive steps for parents.

Real-Life Example – *David and Online Safety*

David was concerned about his 12-year-old daughter's interactions on social media. He knew he had to discuss online safety with respect to sexual relationships but was unsure how to approach it without triggering defensiveness. David chose a relaxed Saturday morning to bring up the subject. He spoke openly about the potential risks and listened to his daughter's thoughts. Together, they set up safety settings on her accounts. The "sex talk" is often a source of anxiety for parents. However, it's essential for children to get accurate information and develop a healthy perspective on sexuality.

Real-Life Example – *Karen and the Sex Talk*

Karen felt awkward discussing sex with her 14-year-old son. She was concerned that bringing it up might encourage curiosity and experimentation. Karen decided to face her discomfort head-on. She researched age-appropriate ways to discuss sex and chose a casual setting to initiate the conversation. To her surprise, her son had many questions and was grateful for the open dialogue.

Discussing topics like sex and grooming with children is a sensitive matter that requires careful consideration, especially when trying to do so in an age-appropriate way. Learning some questions can be an effective way to guide children without shocking them with too much information too soon. Here are some questions you might consider:

For Younger Children (Ages 5–8)

Body Autonomy. *"What parts of your body are private and should not be touched by others?"*

Understanding Consent. *"How do you feel when someone hugs you without asking?"*

Safe and Unsafe Touch. *"Can you tell me about a time when someone touched you and it felt uncomfortable?"*

Trusted Adults. *"Who are the adults you can talk to if something makes you feel uncomfortable?"*

For Tweens (Ages 9–12)

Understanding Boundaries. *"What do you think personal boundaries mean, and why are they important?"*

Friendships and Peer Pressure. *"What would you do if a friend asked you to share a personal photo online?"*

Understanding Secrets. *"Is it okay for someone to ask you to keep a secret from Mum and Dad?"*

Digital Safety. *"Why do you think it's important not to share personal information online?"*

For Teens (Ages 13+)

Healthy Relationships. *"What do you think makes a relationship healthy or unhealthy?"*

Understanding Grooming. *"Have you ever felt like someone was trying to become your friend for the wrong reasons?"*

Sexual Consent. *"What does consent mean to you, and why is it important in any relationship?"*

Personal Integrity. *"How would you handle a situation where someone is pressuring you to do something you're uncomfortable with?"*

Online Behaviour. *"Do you think people behave differently online than in real life? Why might that be dangerous?"*

Time for Reflection

This exercise aims to help parents reflect on the intersection of technology and sex in their children's lives. By taking a moment to think and understand, parents can better guide their children in navigating these complex arenas.

Understanding Your Perspective

- Reflect on your own experiences and knowledge about technology and sex. What was different during your youth compared to now?
- How do you currently view the intersection of technology and sex?

Opening Dialogue with Children

- Think about the conversations you've had with your child regarding technology. How often does the topic of relationships come up?
- Have you ever discussed the concept of consent and boundaries in the digital age with them?

Setting Boundaries

- Reflect on the boundaries you've set around tech use in your home. Do they consider aspects of sex and relationships, especially as your children grow older?
- How do you monitor or oversee your child's online interactions?

Staying Updated

- The digital landscape is ever-evolving. How do you stay updated with the latest trends, apps, or platforms popular among the younger generation?
- How can you proactively educate yourself about potential risks or challenges your child might face online?

Role-Playing for Understanding

- Consider doing a role-playing exercise with your child, where they navigate a hypothetical online scenario related to sex or relationships. How would they handle it? What would their response be?

After reflecting on these points, it might be beneficial to have a sit-down discussion with your child. Sharing your concerns, listening to theirs, and collaborating on solutions can be instrumental in fostering a safe and understanding environment at home. Remember, ongoing dialogue is the key to addressing the challenges posed by technology and intimacy.

In a world where our attention is constantly pulled in different directions, making a conscious effort to truly see and get to know our children is more crucial than ever. While the digital age threatens to make our children's needs and identities invisible behind screens, a proactive approach to parenting can ensure they feel seen, heard, and valued. Staying connected isn't just about managing screen time; it's about enriching the emotional lives of our children and strengthening the parent–child bond. After all, the dream of a harmonious family life in this digital age is one we should all be dreaming together.

Chapter 14

CREATE A PLACE WHERE EVERYONE FEELS THEY BELONG!

Build a Village Where You Are Being Seen

In Brief

In today's world, we're all just a click away from each other, thanks to smartphones and social media. So it's pretty shocking to hear that about one out of every three adults in Australia feels lonely. This isn't just a sad fact; it's a wake-up call. We need to do something about it. Experts like Dr Michelle Lim have been digging deep into why people feel lonely and how it can mess with our health—both in the mind and the body. What she's found is that fixing loneliness isn't just about hanging out with people. It's about really understanding what makes us tick, helping communities come together, and even looking at how society as a whole can make us feel isolated. But here's the good news: with your help, we can roll up our sleeves and tackle this problem from all angles. Your support can help us kick off projects that can truly make a difference in breaking the loneliness cycle. So let's team up and make everyone feel like they're part of something bigger.

The power of friendship and being there for each other is really something special, right? It's also our biggest weapon in beating loneliness. It's about identifying and implementing strategies that can help us reclaim our connections to each other, to nature, and to our authentic selves amidst an increasingly urbanised and individualistic world. Developing strategies to reconnect people is vital for promoting health and well-being in our communities and societies. Remember, the quality of your connections matters more than the quantity, so focus on cultivating deep, meaningful relationships.

Reconnecting and Belonging: We Are a Social Species

The truth is, humans are innately social beings, built more for cooperation than competition. This becomes evident in the mini-communities we call families, which are like microcosms of larger societies—demonstrating our tendency to come together rather than drift apart. In an enthralling conversation with Hugh Mackay, the social researcher and commentator on contemporary Australia, we discussed whether it is time we moved beyond the myth that we're all just out for ourselves, especially when it comes to parenting. The essence of family and community life lies in the mutual support and active involvement of all its members. It's a balancing act between individual freedom and collective responsibility—a dynamic that often starts at home. Parents and children alike struggle with balancing their own needs and wants against the greater good of the family unit. As parents, it's crucial to remember that altruism isn't just a nice idea; it's an integral part of a healthy life. Kindness, compassion, and mutual respect are not just societal ideals but values that we must instil in our children. Teamwork is essential for dealing with life's big challenges. It lays the foundation for effective communication, collaboration, and even conflict resolution.

In an era in which technology can isolate as much as it connects, we have more control than we think, particularly in the digital world where our children now spend a significant amount of time. The transformation in society won't come only from sweeping legislative

Chapter 14: CREATE A PLACE WHERE EVERYONE FEELS THEY BELONG!

changes; it must be deeply rooted in our everyday deeds. These actions, often overlooked, occur in our daily lives—our schools, workplaces, and neighbourhoods. It's here that we lay the foundation for the world we wish to build for our future generations. While it's great that we rally together in times of crisis, why should we wait for catastrophes to act with compassion and support?

In the digital age where "being seen" is often reduced to social media likes and follows, let's redefine the concept. Teach your children that being truly seen involves meaningful connection and presence. Whether it's maintaining eye contact, flashing a warm smile or a good morning to a passer-by, or initiating a conversation in public, these simple acts serve as powerful demonstrations of humanity to our children. Engage proactively in community activities—be it a local book club, a choir, or a community garden project. Show your kids that "being seen" is not about fleeting online engagements but is deeply anchored in real-world, meaningful interactions. The focus shouldn't be on how these actions elevate *your* mood or self-esteem; instead, it should be on the positive impact you're making in *another person's* life.

Ask the important questions: were you fully present when someone needed your attention? Did you apologise genuinely when you erred? Were you forgiving when someone wronged you? Were you there when someone, even a stranger, needed a moment of your time or a word of encouragement? Remember, through the lens of our children, every action we take sends a powerful message, rippling out to shape the ethical framework of the community.

The better future we all desire is the sum of individual efforts, channelled into a unified force for communal progress. This is particularly important as we guide the digital-fluent generations of the future, teaching them that transformative change is not an idealistic notion but a part of daily life. It begins with each one of us, with every conscious decision we make and each meaningful action we commit to.

The Benefits of Peer Interactions for Social Skill Development

Peer interactions and play are essential for our children's development. Lockdowns posed some hurdles with this for many children, taking away peer interaction during critical, formative years. Peer interactions are necessary for children to:

- learn from one another
- understand conversation, emotion, and group norms
- increase independence
- reduce bias and prejudice
- learn about conflict or disagreements (as well as problem-solving skills)
- learn empathy and cooperation.

Peer interaction helps guide our children's social development in groups and teach them what different group dynamics are like. Even simply establishing relationships with other children is a major milestone for the development of the brain. These early peer interactions shape their future relationships in adolescence and adulthood.

Yet, it's important to note that the research also shows that putting two toddlers together won't necessarily mean a friendship match made in heaven. Compatibility still matters at such young ages, meaning children tend to get along well with those who match their cognitive and emotional maturity and play preferences. They are also more likely to seek out play with children who they have had previous positive interactions with.

However, as we know, social support is crucial throughout life. It helps us navigate adversities. It provides us with comfort and connection. It gives us feelings of belonging, safety, and security, which we all crave. While these factors all start at home, they branch out into friendships and other relationships outside of the family circle.

Chapter 14: CREATE A PLACE WHERE EVERYONE FEELS THEY BELONG!

Encouraging your child from an early age to interact with others is critical for their development and future connections. Preschool or play dates can offer the beginnings of learning these social and emotional skills that set us up for a lifetime. Later we will discuss in more detail the effects of play on the brain.

Plan Quality Time With Your Children

While having positive interactions sprinkled throughout the day is wonderful for the development of your child's brain, having time where you give your child your *undivided attention* is also very beneficial.

Now, again, we are all doing our very best with what we know and the time we have. So let this serve as a gentle reminder rather than a harsh rule. If it's available to you, plan special, quality time to spend with your children when nothing else is pulling at your attention. Put your phone away. Halt the household chores for a brief time. This could even mean story-time before bed, attending their sports game, or other small moments.

Spending quality time with our children shows them that we value them and consider their interests important. It can encourage them to pursue these interests and help them build confidence. And, truly, this doesn't take much. Again, small moments work here too. We aren't trying to add more to the already busy parent's schedule. Combining tasks can make quality time with your children both manageable and enjoyable, turning everyday moments into opportunities for connection and growth.

We'll explore the art of mastering quality time not in a hurried, frazzled manner, but with conscious intent. This isn't about rushing through activities just to get them done; it's about weaving threads of intention, connection, and well-being through the fabric of our everyday lives.

Create Time-Rich Rituals

Establish daily or weekly rituals that your family can look forward to. This might be a nightly bedtime story, a weekly game night, or a special weekend breakfast. These rituals provide consistent quality time and create a sense of stability and belonging, strengthening family bonds over time.

> ### Real-Life Example – *"Tuesdays for Two"*
>
> Tuesday mornings. No matter what was going on, Tuesday mornings would be untouchable. It was to be their sacred time, a standing breakfast date with her daughter Tracey. The rule was strict: it would not be rescheduled or skipped for any reason, except for emergencies.
>
> "Don't forget, breakfast with me on Tuesday before school. It's a date!" Donna reminded Tracey on Sunday evening.
>
> "Sounds fun, Mum," Tracey responded with a smile, genuinely pleased.
>
> The first Tuesday, Donna was already up before the alarm went off. She quickly made herself presentable and drove to Tracey's favourite breakfast spot—a cosy little café known for its avo on toast and freshly brewed kombucha.
>
> Over breakfast, they talked. Really talked. About school, Tracey's friends, and even the boy she had a crush on. Donna listened, gave advice when asked, and sometimes simply sat in comfortable silence. It wasn't just the food that was nourishing them; it was the conversation, the laughter, the uninterrupted time to connect.
>
> Weeks turned into months, and Tuesday mornings became an institution neither was willing to skip. Tracey began

opening up more, sharing her concerns about schoolwork, her dreams for the future, and her thoughts on everything from politics to pop culture. Donna felt like she was rediscovering her daughter, seeing the young adult Tracey was becoming. She found herself looking forward to Tuesdays more than anything else in the week. Work could wait; errands could be rescheduled. Those precious Tuesday breakfasts became her sanctuary, a space where she was just "Mum", cherishing the company of her growing daughter.

One Tuesday, as they were leaving the café, Tracey hugged Donna tightly and whispered, "Thanks, Mum. I love our Tuesday mornings."

Donna hugged her back, eyes moist, and thought, *Me too, sweetheart, me too.*

In a world that was always demanding more—more time, more attention, more everything—Donna and Tracey found that the best "more" was more time with each other. And all it took was a plate of avo on toast, two glasses of kombucha, and the promise of Tuesday mornings, a time reserved exclusively for them.

Integrate, Don't Isolate

The first strategy for finding quality time with your children is to integrate them into activities you're already doing. This could be cooking, gardening, exercising, or even grocery shopping. Not only does this allow you to spend time together, but it also provides a space for your children to learn practical skills, understand responsibilities, and appreciate the value of everyday tasks.

One powerful method is the act of playing with your children, especially in nature. Yes, this means you can achieve your daily dose of physical exercise while bonding with your little ones. By bringing them into your routine, not only do you check off your wellness box, but you also set a strong example for them. When they see you valuing physical activity, they mirror this behaviour, setting the groundwork for lifelong habits of movement. By incorporating nature into these shared activities, you can further enrich the experience. Whether it's a jog in the park, a game of catch in the yard, or a yoga session under the morning sky, the added layer of nature's tranquillity can enhance your mood, promote better focus, and provide a refreshing environment for you and your children.

Double Up on Benefits

Look for activities that offer multiple benefits. For instance, a walk or bike ride in the park can be a moment of physical exercise, an opportunity to connect with nature, a space for meaningful conversation, and a break from digital distractions. In later chapters, we'll tackle how smart dietary choices coupled with regular movement can help regulate appetite—for both you and your children. The key is to identify activities that align with your family's needs and values, then leverage them for multiple purposes.

Plan for Downtime

In our busy lives, it's easy to overlook the value of simply doing nothing together. Planning for downtime—be it cuddling on the couch, stargazing, or enjoying a lazy Sunday morning—provides space for relaxation, open-ended conversations, and a shared appreciation for quiet moments.

In combining these strategies, remember that quality time is less about the quantity of time spent and more about the presence, engagement, and intentionality you bring into these shared moments. Your aim is to cultivate a life that allows your children to thrive, creating a web of shared experiences that help shape them into well-rounded individuals.

Chapter 14: CREATE A PLACE WHERE EVERYONE FEELS THEY BELONG!

Each thread we explore in this chapter—from shared physical activities and nature engagement, to eating together—will gather into a strong, colourful ball. This isn't just any ball, though; it's an interactive tool, a playful metaphor that you can "bounce around" in your day-to-day life. It's a reminder that time isn't just something to manage—it's something to dance with. By mastering time you are set to become a thriving parent who raises thriving children, turning each day into a symphony of well-being, connection, and joy.

Reconnecting With Your Child

Love can be passed down through generations. Similarly, a lack of love can also be passed down. Remember Dr Michael J. Meaney's findings with the baby rats who were taken from their mothers and never experienced their mothers' grooming? They secreted more stress hormones than baby rats that did received this affection and care. Meaney and his team further found that when this affection was missing, the offspring failed to show *their* babies affection as well. It becomes a vicious cycle. But this cycle can be broken.

While we've already explored the importance of a serve and return relationship, the following can offer even more within this relationship. The tips below can provide a gentle nudge in the right direction for creating healthy and positive interactions with your children that will benefit both of you.

Read Together

Those bedtime stories. They might mean more than we think! Dr Mark Williams has shown that reading on paper is one of the most important things we can do to learn, feel more confident, and promote well-being across our lifespans. Encouraging our children to read, no matter whether they're reading a comic or magazine, has profound impacts. The key is to get them to love doing it! Another thing to note here: research has shown that reading specifically on paper is important. At the end of the day, education is one of the most critical factors for generating lifelong health and well-being, and we can encourage our children to love learning and reading early on.

Learn to Laugh More

Countless studies indicate the positive impact of laughter on our cognitive function, memory, and overall well-being. When we laugh, we activate the "joy" centres of the brain. Interestingly, dancing and laughing produce similar effects. They both offer ways of experiencing joy and dampening stress, forgetting the worries of the day and immersing ourselves in the current interaction and the happiness it brings. And we have the power to tell a joke or dance more!

Research even shows that *shared laughter* is a sign of a happy relationship. Laughing can strengthen our bonds with others and enhance the love between us. Plus, we all know how contagious laughter can be. Sometimes it's hard to stop when someone close to us is in a fit of it. And we know it feels good!

So, how can you laugh more in your life and with your children?

- Play together! An imaginary world can bring about various shocking (and laughter-inducing) situations.
- Hike, bike, swim, and engage in activities where you can have fun together!
- Cook together.
- Shop together.
- Do finances together.
- Learn a language together.
- Travel together.
- Visit grandparents and aunties and/or learn about family history together.
- Tell a joke or two, and encourage your children to do the same.

Chapter 14: CREATE A PLACE WHERE EVERYONE FEELS THEY BELONG!

- Watch a funny movie together.
- Try not to take life too seriously.
- Trouble laughing? Start with a smile.

Remember those mirror neurons? Smiling activates the mirror neuron system in our brain. This means if we see someone else smile, we are more likely to smile back, and vice versa. If you struggle to laugh, force a smile until it becomes more natural. This will activate the pleasure centres in your brain. You may even want to consider smiling at yourself in the mirror each morning to make this more second nature. And remember, our children often do as we do; smile more, and they will likely do the same!

Prioritise and Normalise Parent–Child Physical Affection

For those of us who may not have received physical affection as a child, this can feel strange at first. But with everything, practice and repetitive actions are key.

Start with a smile, as per above. Then, try simply holding your child or partner's hand. Once this becomes comfortable, try a hug. Hugs can truly heal us. We all are sensory beings with thousands of nervous connections that require stimulation. You may even notice your own insecurities and shyness vanish within seconds of a hug.

From here, the key is to incorporate these bits of physical affection throughout your interactions with your child to foster love and connection. Here are a couple of other tips you can use:

- **Make Time for Cuddles.** When watching a movie or reading a book, encourage closeness within your family unit. This could also mean carrying your child when they become too tired, or tickling or stroking. This will help them feel safe, relaxed, secure, and confident, and strengthen your bond while nurturing good brain health.

- **Incorporate Affection into Everyday Activities.** Bath time, nappy changes, and getting dressed or ready for the day all offer opportunities to connect. Or, while out and about, you can point at things that you think your child might be interested in, showing them the attention that they very much need. These small interactions don't add time or busyness to your schedule but fit right into what you're already doing.

Try to Give Your Child Unconditional Love

It's easy to love our children when they are happy and easy-going. It's harder to love our children when they are maybe screaming at the top of their lungs or putting all their energy into fighting us to get dressed. But love shouldn't come with conditions.

Knowing we have a safe and secure person or place to turn to in the worst of times can provide us with resilience to stress and help guide us through these inevitable dips that happen as we journey through life. Sometimes, a simple hug or touch on the shoulder or an "I'm here" is all we need to push through. Real love doesn't need a wall or a fence. It knows no boundaries.

A common mistake many parents make is trying to force their children into being people they're not. Instead, it's much better that we love and accept our children for who they are. They might not excel in the sport we hoped they would, but they have strengths and skills in other areas that we can appreciate and encourage. Love is acceptance without barriers. And when we give this kind of love, our children will pass it on, creating a rippling and positive effect elsewhere, potentially in places we never imagined.

Forgive Yourself (and Others)

Parent guilt is an all-too-common shadow within society. As parents, we may feel like we are never doing enough or spending enough time with our children. However, learning to forgive ourselves and

Chapter 14: CREATE A PLACE WHERE EVERYONE FEELS THEY BELONG!

extending that forgiveness to others offers flexibility and grace, especially during difficult times.

Practising forgiveness is shown to produce positive brain states. It activates areas of the brain associated with empathy and emotional regulation. It has the power to dissipate anger and resentment and create understanding within us and with others as opposed to building walls. None of us is perfect. Striving to be will only result in disappointment. Instead, expect mistakes. Expect to learn from them. Forgive yourself for making them and know you're only human.

An important distinction to make here is that forgiveness doesn't mean forgetting. Rather, it's finding a way to move past negative emotions and move forward. When we demonstrate forgiveness and compassion with ourselves and our children, we teach them how to cultivate these in their own life. We teach them it's okay to make mistakes and self-correct. We show them resilience and healthy ways to cope. And yes, we can thank mirror neurons again here!

Time for Reflection

Reflecting on our social connections can give us valuable insights into our current emotional state and well-being. By understanding our social ties we are able to design proactive steps to fortify them. Consider all of the following areas of your life and how you might manage them.

Your Inner Circle

- List five people you're close to.
- Plan a face-to-face meet-up with at least two of them in the coming weeks.
- Aim for a meaningful conversation. Maybe set a tech-free time or choose a serene spot for your meet-up.

Digital World vs Real World

- Track a day's social media use. Aim to reduce it by 10% the next day.
- After each online session, take a five-minute break to stretch, breathe, or have a quick chat with someone.
- Try to meet a friend in person once a week or fortnight.

Loneliness vs Solitude

- Recognise your feelings in crowded spaces. If you feel lonely, consider joining a club or group that aligns with your interests.
- Enjoy your alone time. Schedule a regular "me time"—it could be reading, walking, or just relaxing.

Your Support Network

- Identify your support figures.
- Initiate a regular meet-up: perhaps a monthly dinner with family or a bi-weekly coffee with a mentor.
- Attend community events or workshops. It's a great way to meet support figures and also expand your circle.

Workplace and Community Bonds

- Engage in team-building activities or casual meet-ups after work.
- Reflect on your work–life balance. If work is isolating you, consider flexible hours or remote working days for a change.

Finally, based on your reflections and these action steps, write down a "social prescription". This plan will guide you to cultivate richer relationships and a deeper sense of community belonging. Remember, every little step can make a significant difference.

Chapter 14: CREATE A PLACE WHERE EVERYONE FEELS THEY BELONG!

If you take anything at all away from this chapter, it should be that love and connection precede all else. At the heart of good health is unconditional love, affection, and connection, as they positively shape us and our abilities to interact with the world. When we build positive and loving relationships with ourselves, our children and others, we aren't only creating closeness for many years to come, but we are also guiding them on the basics of what healthy relationships look like and how to interact with others respectfully.

Beyond relationships, stress, and love, our food and how much we move also play a significant role. In the next chapter, we dig into how the food we eat influences our and our children's brain health.

Chapter 15

LIVING A VITAL LIFE

The Power of Whole Foods and Fitness

In Brief

Imagine a dinner table on which your family's plates are alive with vibrant colours, delightful flavours, and wholesome foods. These natural and untouched foods are nature's gift, giving our kids the best start every day. But our quest for our family's well-being goes beyond the kitchen. Combining these meals with active family moments creates a winning combination. It could be an evening walk with the kids, a weekend bike adventure, or just a spontaneous dance-off in your living room. Movement complements a whole-foods diet, making every benefit even more powerful. But here's the thing: as parents, our lifestyle choices don't just affect us; they shape our children's futures too.

Being a healthy and fit parent is leading by example. When our kids see us making choices that prioritise our health, they're more inclined to follow in our footsteps. After all, children often mimic their parents. If they see you enjoying a fresh salad or heading out for a morning jog, they're likely to adopt these habits as they grow up. The added energy and vitality we gain from a balanced diet and active lifestyle also mean we're better equipped to handle the demands of parenthood, be it playing soccer in the backyard or going on family hikes. But it's not just about setting an example; it's about creating an environment where health and happiness thrive. In this chapter, we'll explore how our choices as parents can pave the way for a brighter, healthier future for our children. We'll delve into the benefits of whole foods, the joys of an active lifestyle, and the ripple effect these can have on our family's well-being.

What's in Your Grocery Basket

What we pick up during our supermarket visits says a lot about our health aspirations and current cravings. Each item in our grocery basket, whether it's fresh produce or a packaged snack, affects our health. Are we choosing health or convenience? Taste over nutrition? Let's unpack our grocery decisions and discover the deeper tales they tell. Often, the freshest and healthiest options lie on the store's perimeter—from fruits and veggies to dairy and meats—while the centre aisles lure us with packaged conveniences.

By now, if I've done my job right, you likely understand the problem with processed foods. Understanding that processed food has very little nutrition, lots of sugar, no fibre, and no nutrients is the first step towards making appropriate changes. So, how can we expand our awareness and apply this knowledge to our everyday lives? We certainly don't want our children to succumb to the life-threatening and debilitating diseases caused by high consumption of sugary and processed foods.

The good news is that reducing our sugar intake doesn't need to be difficult. A great place to start is by simply looking at the labels of the food you eat. For example, strawberry low-fat yoghurt, tomato sauce, and other items are full of sugar. I encourage you to check the labels; you'll be surprised how much sugar is hidden in our daily foods.

At the same time, we don't need to do anything drastic here, such as overhauling our entire pantry (although, if you feel this is right for

Chapter 15: LIVING A VITAL LIFE

you, by all means …). We can start by taking simple steps to swap our processed foods for healthier whole-food items. So, let's dig into the action and discover easy tips and tricks to help your family eat healthier—and, in turn, elevate your and your child's brain health.

Easy Tips for Healthy Eating

When I decided to take charge of my own health, I overhauled my diet by removing sugar, processed food and incorporating nutritious whole foods like vegetables and fruit. The benefits extended beyond just physical health. As I incorporated stress-management techniques, like pausing the smoke detector discussed previously, there was a noticeable uplift in my mood, energy levels, and overall confidence. A delightful bonus was the re-emergence of my waistline! My journey wasn't just about looking good; it was about feeling good from the inside out.

And that's my aspiration for you. I want you to experience the power of nourishing food and beverage choices. I hope that you, your family, and future generations will understand and appreciate the deep connection between what we consume and how we feel. A healthier, happier, and more vibrant life is within reach, and it all begins with what's on your plate.

When we sit back and think about when we consume processed food or sugary foods and drinks, we probably notice that it tends to be when we are tired, stressed, lonely, depressed, or angry. Then, when we grab our favourite snack or drink, we feel better for a few minutes. It makes us feel good. It's not necessarily sugar for all people, either; sometimes, it's salty or fatty foods.

Either way, these sugary and fatty foods trigger the same reward areas of the brain as alcohol and nicotine. But how much sugar is too much sugar? Well, according to the World Health Organization (WHO), men shouldn't be drinking or eating over nine teaspoons a day, women over six teaspoons a day, and children over five.

Most of us are eating and drinking well over this recommended amount. Added sugar sits at the heart of Australia's worsening health crisis, with teenagers consuming an alarming average of more than 20 teaspoons a day, surpassing the national average of 16 teaspoons. The World Health Organization (WHO) advises that added sugars should make up no more than 10% of our daily caloric intake, and that lowering this to 5% would be even better. Sugary drinks are the primary culprits in the surge of added sugar consumption, particularly among the youth. So, here's how we can cut back!

STEP 1: Pay Attention to What You and Your Family Are Eating

We tend to be of the mindset that if foods and drinks are available to purchase in a shop, they must be healthy. They are often marketed as healthy. Thus, you may be led to believe you are doing the right thing for your children by rewarding them with a healthy treat after their soccer game or homework. But let's take some time to pause when purchasing food.

What does the label say? How much sugar does the item contain? Checking the fine print before placing the item in our cart can help us make informed and better choices, as well as cut down on our family's sugar consumption.

If you want to take this a step further, I encourage you to track your sugar intake over one week. Many phone applications can help you do this (when in doubt, there is also Excel or Google Sheets or a good old-fashioned notepad). This can be eye-opening and make you more aware of what you're putting in your body and feeding your family. Awareness is always step one!

STEP 2: Simplify Your Grocery Choices

It's all very well to check the labels when we're in the supermarket, but beware of marketing tactics. It's easy to get swayed by them, especially when we're juggling work, parenting, and personal commitments. Is that product really "healthy", "natural," or "fortified", as the advertising claims? For time-poor parents, here are some streamlined strategies:

- **Create a Go-To List.** List down your family's *favourite* healthy foods and stick to them. It reduces decision-making time and ensures you're consistently picking nutritious options.

- **Identify Sugary Culprits.** Recognise that sugary beverages are a major source of added sugar, especially for young people. Be aware that a standard 600 ml cola bottle may contain about 65 g or 16 teaspoons of sugar, which is over twice the recommended daily intake of added sugar. *If you can only do one thing, start here!*

- **Use Technology to Your Advantage.** Instead of manually checking each label, there are apps available that can scan barcodes and instantly provide nutritional breakdowns. These apps can be a lifesaver, highlighting sugar content, additives, and more. *Use online shopping to avoid buying from the middle of the grocery store.*

- **Limit the Number of Processed Foods.** Stick to the perimeters of the grocery store where fresh produce, dairy, meats, and seafood are usually located. This will naturally reduce the temptation to pick up processed foods.

- **Schedule Regular Grocery Trips.** Setting aside a specific day and time each week for grocery shopping can help ensure you're stocked up on healthy options, reducing the chances of last-minute unhealthy food purchases.

- **Educate and Involve Kids.** Turn grocery shopping into a mini-lesson. Teach your children to spot high-sugar items and challenge them to find healthier alternatives. This way, they become more conscious of their choices and understand the importance of healthy eating.

- **Set a Goal for the Week.** For instance, try reducing the number of sugary snacks you buy each week. Having a tangible target can motivate you to stick to better choices.

Being aware is indeed the first step, but integrating these straightforward strategies can ensure that step becomes a leap towards healthier eating habits for your family. Remember, every small choice accumulates into a broader impact on your family's well-being.

STEP 3: Take Out One Fatty or Sugary Food at a Time

I was quite shocked when I realised how much sugar I used to consume. And while removing sugar sounds simple, it's not easy to do. I started with cutting out after-dinner chocolates for a week; after mastering this, I nixed the trail mix, which was my go-to after stressful work meetings. It's important to make this a process to avoid overwhelm, more stress, and the inevitable frustration that may follow.

Because eating sugary and fatty foods relieves the immediate feelings of stress, it's important to think about what to replace them with to achieve the same effect. In my case, I turned to raw nuts.

You might discover major benefits after reducing your sugar intake; you might start to feel full after meals, your waistline might shrink, and your days of going through seesaw diets might disappear.

You might also benefit from incorporating stress-relief techniques into your daily life. Ask yourself when reaching for food: am I hungry—or am I stressed? If you notice you're stressed, you may benefit from turning to deep breathing techniques or exercises. At the end of the day, finding what works for you is key to keeping your stress levels down and avoiding turning to junk food as a means of relief.

STEP 4: Incorporate More Whole Foods and Fibre into Your Diet

Whenever we remove something from our lives, it creates a vacuum. In other words, other habits or things will fill up the gap if we don't

Chapter 15: LIVING A VITAL LIFE

intentionally do this ourselves. While stress-relieving activities may help you cope better, there's still the matter of what to eat.

Instead of solely focusing on *removing* sugary items, steer your efforts into incorporating more whole foods, such as vegetables, meats, fruits, dairy products, and more. Basically, anything that comes out of the ground or that is an animal product is best. However, there are a few things to note here as well.

When the food industry processes these foods, they may change the nutritional profile or even add unhealthy chemicals like emulsifiers, additives, and more. Now, I don't say this to add more fear into our already fearful world. Rather, it's to give you another level of awareness, simply to help you do the best you can with what you know.

Take a chicken breast, for example. Where was the chicken raised? What did it eat? Surprisingly, there is a difference between factory-farmed and pasture-raised or free-range chicken. A factory-raised chicken eats corn, which is not a food staple for any species. There are no wild animals that eat corn routinely. Chickens and cows are fed only corn in concentrated animal feeding operations because corn is higher in energy than virtually any other food. It's also very affordable for those individuals or corporations raising chickens for the general population's consumption.

While an animal raised free-range and not fed corn is much better for us to eat, it is very expensive. Then, unfortunately, there's the fact that many of these chicken breasts are soaked in a salt, sugar, and water solution so they swell, and they can, thus, command a higher price at the cash register. Again, not ideal. But we can do the best we can by trying to purchase meat that is grass-fed or free-range over factory-farmed or pasture-raised.

Again, reading and understanding the labels on the food we eat can go a long way towards helping us make the better choice. And again, when it comes down to it, the research shows over and over again that whole foods trump processed foods every time.

Simple Tips for Getting Kids to Eat Their Veggies

When my children were very young, as much as I wanted them to eat their veggies, no amount of negotiating or persuasion could get them to eat certain foods (and oh, how I tried!). Sometimes, I would just let them have it their way. It felt near impossible figuring out how to eat "healthy" as a family. So, if you're feeling this way, know it's not just you. And on top of this, if you're concerned—well, these are the markings of a parent who truly cares. Rest assured that you are doing something right.

While we know that giving our children a variety of foods early on (and even eating a variety of foods during pregnancy) may help reduce the odds of having a picky eater, kids can change their minds about what they like, or the look of the food might motivate them to refuse it entirely.

Being a parent myself and talking with countless other parents, I pulled together a few tips that might help, even on those very busy days.

- **Hide Vegetables.** This doesn't mean tucking them away in cupboards or the back corners of the fridge. Instead, try hiding vegetables in foods or meals your children already enjoy and love. For example, many muffin recipes offer easy ways to hide grated carrots or zucchini. Vegetables can also be pureed and added to sauces, soups, smoothies, and more.

- **Keep It Simple.** Roasting vegetables with meat for dinner in the oven or on the BBQ can offer a quick way to make a healthy meal. Then, allow each family member to add some of their favourite sauce once it's dished out.

- **Have Easy Snacks Available.** Some examples include yoghurt and fruit or nuts, vegetables and dip, fruit and dip, sandwiches, and smoothies.

- **Set the Example.** We know kids do as we do. We can provide a good example by showing them how enjoyable vegetables can be!

Chapter 15: LIVING A VITAL LIFE

You Can and Should Tell Sugar and Ultra Processed Food to Get Lost!

The best place to start on reducing your sugar intake is reading the packets of processed foods and sugary beverages. Eventually this will become a long-term habit. Your awareness of sugar will increase, and you'll be more inclined to make better choices.

As you train your brain to say no to sugar and ultraprocessed foods, you will boost your brain health and overall health. As parents and caregivers, we can influence our children to do the same. So, the next time you find yourself reaching for that treat or drink, take a moment and imagine saying, with a cheeky grin, "Sorry, not today!" Envision your brain cheering you on, thankful for the healthy choices you're making. Picture your children's futures bright and filled with potential as they too learn to make wiser food choices. Remember, every time you opt for healthier alternatives, you're not just choosing for yourself, but for the generations to come. After all, there's no sweeter victory than living a healthy, thriving life. Know that you are in charge!

The Mood-Boosting Benefits of Fibre

Diet and mood are intricately connected. While we often hear about sugar crashes or the euphoria from chocolate, there's a dietary hero in the mix: fibre. This humble nutrient, found abundantly in fruits, vegetables, and grains, has profound benefits for our gut health. But how does this relate to our mood? The gut is often referred to as the "second brain", and the health of our gut can directly influence our mental well-being. Consuming a fibre-rich diet promotes the growth of beneficial gut bacteria. These "friendly" bacteria produce compounds that positively affect brain function and mood. Moreover, a balanced gut can help regulate blood sugar levels, preventing mood swings. So, the next time you're feeling a bit down, ask yourself whether you've had your dose of fibre. A happy gut can often lead to a happier you.

Leading by example is crucial, and it's more than just what we say—it's what we *do*. When we prioritise nutritious foods and incorporate regular movement into our routines, we not only benefit ourselves but also provide a practical blueprint for our children. But what exactly constitutes "regular exercise"? I once thought training for a marathon would be my ticket to fitness. However, despite the miles I clocked, I noticed little change on the scales. Puzzled, I realised that occasional spurts of activity weren't enough to offset prolonged periods of inactivity. It wasn't just about the intensity of the exercise but also the consistency. It was about counteracting sedentary habits. Thus began my journey to movement. Let's explore what regular exercise truly means and how to integrate it seamlessly into our lives.

Time for Reflection

Here's a quick and engaging way to think about family food choices. Let's make our meals more colourful!

Our Food Choices
- Chat about yesterday's meals.
- Find foods in those meals that contained fibre, like fruits, veggies, or grains. How many were there?
- Idea for tomorrow: How about adding some berries to breakfast?

Fibre Fun Time
- Pick three fibre-filled foods everyone likes.
- Try to eat at least one of them every day this week.
- Tip: kids can sprinkle seeds on their cereal or choose fruit for a snack.

Chapter 16

TAKE FORWARD ACTION

Moving and playing

In Brief

Forty years ago, after she had raised four kids and weathered the many storms that life throws at you, my mother's health deteriorated. To take back control, she started walking. Back then, this was quite unusual, and she would often be the only one out during the early morning hours. When the weather took a turn for the worse, she would walk around the pool table at home. One of my kids' favourite memories is sitting on top of the pool table as Behar, as they call her, walked circles around them.

For my mother, walking has become the best medicine. Compared to others her age, she's in incredible shape.

Meanwhile, I kept coming up with all the reasons why I had zero time to exercise, which I'm sure many parents can relate to. I'd come home from work, get the kids dinner, put them to bed, then night after night, find myself falling asleep on the couch in my work clothes. As with most parents, time was not on my side.

I remember my mum calling me on my fortieth birthday. The conversation started off with the usual "Happy birthday!" and "Did you have a lovely day?" Then, my mother said something that caught me off guard: "Well, now that you've turned forty, you have to start exercising."

Of course, my answer was anything but receptive. Exercise? When? At 3:30 am before work and before the kids were up? At 10 o'clock at night, once everyone went to bed?

The conversation didn't entirely end there. My mother would check in with phone calls asking me if I was exercising. And as much as we fight our own parents on these wise words, she was right. I needed to be moving my body more than I was. But I had to become unhealthy and gain weight before I understood this. The good news is that it's never too late to start.

We all know exercise is good for us; actually doing it, on the other hand, is a whole other story. The good news is that exercise isn't the whole story when it comes to better brain health. And there are also many ways to move the body, and most experts indicate that finding an activity you enjoy is the most important thing.

Let's Look at the Neuroscience Connection

When it comes to the developing brain, movement provides physical benefits but also the freedom to make mistakes, opportunities for teamwork, and the development of numerous other skills. And one country has learned the benefits of combining music, art, exercise, and team sports for youth far faster than others. Almost two decades ago, Iceland began implementing a program to help deter teens and youths from drug and alcohol use (which, as we've learned in previous chapters, is more prevalent in those with extreme ACEs). In fact, where it used to be normal for teenagers to roam the streets in a drunken stupor, and about 42% of teens reported drinking at least once in the past month, now only about 5% claim to have had one drink in the last month. Today, Iceland claims the top spot for having the cleanest-living teens, a promising development for good brain health and mental wellness.

So, how'd they do it? The introduction of after-school classes revolving around music, art, dance, martial arts, exercise, team sports, and anything else the teens were interested in provided much-needed stimulation while also keeping them away from alcohol and drugs.

All of this is to say that the needs of the developing brain and a healthy brain go beyond mere physical activity. There's another huge aspect we have to consider: play. Even as parents, we can benefit from play, which offers a great opportunity to move. So, in the following sections, we'll take a closer look at the impact of play and physical activity on the brain, as well as actionable tips and tricks to help your family play and move more, without the need to get up at 3:30 am or stay up past 10 pm just to squeeze in a workout.

The Impact of Play and Physical Activity

Humans aren't the only species that "play". This activity is seen in various other mammals, indicating an evolutionary drive or need. Research even indicates that an absence of play may lead to various psychiatric disorders. When deprived of this essential developmental guide, individuals may further experience impaired social, emotional, and cognitive impairments as adults.

But here's the key piece: play isn't an activity that revolves around rules, referees, or coaches. While sports and other physical activities can include aspects of play, there's also a need for the "rumble-and-tumble" type of play, such as building a sandcastle from one's imagination or "playing" out a make-believe scenario. This is the same reason many animals "play fight" when they are only weeks or months old. As briefly mentioned above, humans and animals have a lot in common in this regard. In many species, play involves taking turns or playing fair in some shape or form.

In many ways, play is a way for the developing brain to gain invaluable experiences that guide the structure and function of various parts of the brain. These interactions set the stage for future social interactions, relationships, and dynamics. They teach us how to connect with others in positive ways. They may even have the power to turn certain genes on or off.

Between parent or carer and child, play also offers a pathway for connection and love. Play may also help the adult brain maintain flexibility and be a major focal point of joy and fulfilment. It releases endorphins, which help combat stress and make us feel good. So, let's dig a little deeper here and uncover the incredible ways play shapes and changes the brain.

How Play Changes the Brain

Put simply, play changes the neural connections in the brain. It can do this because of neuroplasticity. This is powerful, particularly during those first years or two decades of life.

In terms of brain development, play helps:

- improve executive function
- regulate emotions
- improve problem-solving skills
- enhance the ability to plan
- lay the groundwork for love and relationships.

In studies on rats, researchers have further uncovered that play increased the Brain-Derived Neurotrophic Factor (BDNF) in regions of the brain associated with emotional regulation. BDNF is a protein that's crucial for developing new synapses. BDNF further plays an important role in proper brain development and the formation of neural connections. In fact, disruptions to the role that BDNF plays in brain development have been shown to lead to cognitive function disorders and other detrimental impacts.

It's Time to Play More!

Instead of focusing on finding time to hit the gym every day (a near impossibility for many parents), we can incorporate movement in other ways, such as by playing with our children. This is a win–win. Our children get much-needed movement, attention, and brain stimulation. We bond over the experience. Our children feel seen, get attention, and feel a sense of belonging. And we also end up moving our bodies more. In most cases, we can allow our children to lead the play activity. At other times, we might want to take the lead, such as by setting up an obstacle course for the whole family or by planning a scavenger hunt around our property.

As our children grow, we can also encourage physical activity and play as a family, by, for example:

- going on bike rides together
- going on after-dinner walks
- hiking in nature
- playing sports together
- holding friendly competitions, such as who can hold a plank the longest
- dancing
- jumping rope
- playing at the park.

The Most Dangerous Thing We Do Is Sit All Day

As parents, we want to ensure we're setting a good example for our kids. When we eat better and move more, this reduces our appetite, making weight maintenance and loss that much easier. It also shows our children how to maintain a healthy lifestyle.

To get myself moving, I set myself an ambitious goal to complete a marathon. Given my fitness level back then, it appeared an unattainable feat. However, it was this very challenge that fuelled my daily running sessions. A seasoned marathoner friend recommended the Napa Valley Trail Marathon, describing it as a predominantly downhill run. But the trail, winding through the picturesque vineyards along the Silverado Trail, proved otherwise.

I gradually increased my running distance. Despite the rigorous exercise, my weight remained stagnant. Delving into research, I discovered that prolonged periods of sitting could negate the benefits of aerobic activities. Determined to change, I transitioned to a standing desk, both at work and home. The immediate switch to standing all day initially left me drained. But the benefits soon became evident—better thermal regulation, reduced appetite, and enhanced cognitive clarity. More importantly, it complemented my marathon training, and I started shedding weight.

To strike a balance between standing and sitting, use an adjustable desk. If you find constant standing challenging, it's imperative to include movement into your routine. Exercise is not just a means to shed weight; it's a powerful tool to revitalise ourselves. The connection between good health and physical activity is undeniable. Thus, the challenge lies not in understanding its importance, but in initiating and maintaining an active lifestyle. Taking on this mix of standing and moving around not only made me feel better, but it also helped me be there for my kids with more energy and focus.

What Does This Mean for Busy Parents?

The average person sits for 7.7 hours per day. While getting to the gym at least three times a week would be wonderful, it's not realistic for everybody, especially the busy parent. Studies even show that exercise alone won't help us lose weight. So, what gives?

When the problem seems to lie in sitting for far too long, we can find ways to counteract this by avoiding sitting as much as we can. As previously mentioned, we can play or get active with our family unit as opposed to spending countless hours sitting in front of the TV. We can:

- alternate between standing and sitting as we work
- take frequent walking breaks around our home or office
- walk whenever possible, even parking farther back in the parking lot
- assess how much we sit and determine ways that work for us and our lives to counteract this.

So, perhaps the goal, especially as our children grow, isn't to exercise X number of minutes a week, but instead to simply focus on sitting less. This is attainable and realistic for most of us and won't require a complete overhaul of our routine. Research even demonstrates the importance of moving our bodies during pregnancy. In fact, studies show that children of women who moved regularly during their pregnancy experienced improved cognitive performance, indicating its significance in brain development. Meanwhile, a sedentary lifestyle can lead to brain degeneration and even the eventual development of Alzheimer's disease, multiple sclerosis, and diabetes. So, let's take the pressure off a bit. It'd be incredible if we could incorporate a gym or exercise routine into our lives, but with no time, this simply isn't possible for many parents. Instead, let's find ways, such as using a standing desk, to sit less and move more in our already existing routine.

Besides, our health and our children's health needs go beyond planned exercise. To the frustration of many gym owners, countless clients commit to a regular workout routine but achieve little progress. If these individuals are sitting all day or eating primarily sugary or processed foods, they may very well be undoing what they do at the gym. As more and more research emerges it seems we need to be moving our bodies periodically throughout our days, not merely committing an hour three to four days a week at the gym.

The sad news is that if our parents are overweight or obese, we have a much higher likelihood of also being overweight or obese. The good news is that, as parents, we can change this cycle. Again, our children do as we do. If we focus on setting the example, our children naturally follow. They look to us to learn how to live and do life.

When we focus on improving our health and movement (not obsessively, though, as life should never primarily be about wellness, and with various health advice out there, it easily can become so), we pass this wisdom down to our children. We can commit to sitting less and simply moving our bodies when we find those opportunities in our routine, such as taking a work meeting while walking or getting water every hour to make us stand up.

Embrace the Momentum: You Can Begin Now

Consider the incredible journey of Charles Eugster, a British dentist often hailed as "the world's fittest 90-year-old". At 80, disheartened by his deteriorating physique, he ventured into running, bodybuilding, and rowing. Remarkably, at 95, he set a record in the 200 m sprint for the British Masters Athletics Federation.

Reflecting on his life, he says, "It was only in my forties, when my health was deteriorating, that I truly grasped the wisdom in my mother's advice. The initial step is always challenging, and for me, it involved unlearning years of physical inactivity."

All Our Experiences Shape Our Brain

Our experiences literally shape the brain. Learning a second language, travel, and other experiences all have the same thing in common: they take advantage of neuroplasticity.

How Learning a Second Language Transforms the Brain

Research indicates that learning a second language is linked to a better attention-switching ability, improved working memory, and increased functional connectivity. In turn, it's thought that practices such as learning a new language may offer protective effects against dementia and neurodegenerative diseases.

Chapter 16: TAKE FORWARD ACTION

When we learn a new language, our brain adapts. Over time, new neural networks are formed, including those associated with recall, memory, and creativity. In fact, learning a new language can even increase the size of the brain. And you'll notice the improved function happening if you've ever learned a new language. At first, things will be difficult, but soon you start slowly picking up more and more words and phrases. You begin understanding more and being able to communicate more. As time goes on and you continue down this path, it will become second nature, and you will become fluent in the new language.

The Impact of Travel on the Brain: The Good vs the Bad

Travel offers a wonderful way to expand the mind and flip our perspective, as well as provide us with an abundance of new experiences that we wouldn't get otherwise. Many who travel come across as open-minded and flexible individuals. And there's a reason for this!

Novelty, or newness, activates the brain's reward system and dopamine. While more research is coming out every day, scientists note that there's a connection between this dopamine hit and learning. When researchers inhibited the dopamine receptors in rats, their ability to learn slowed. This suggests that novelty may help us enhance neuroplasticity and keep the brain healthy and happy.

And travel is a wonderful way to experience novelty. When we experience new cultures, new food, new rituals, new adventures, and more, the brain adapts. It rewires based on these experiences. It also gets a healthy dose of dopamine as opposed to unhealthy doses that may come from drugs, processed foods, or alcohol.

But there is a downside to travel. When travelling across multiple time zones, jet lag can be difficult to shake. For many, it takes days or even weeks to adjust, with some experiencing gastrointestinal issues, headaches, irritability, poor concentration, cognitive deficits, and sleep difficulties. While often these effects are temporary, frequent travel across different time zones can have more lasting impacts, causing the brain structure and function to suffer. Thus, some travel can be enriching, but too much, too often, could be detrimental to our brain health and happiness.

At the same time, there's lots of evidence supporting the importance of new and enriching experiences. Aside from travel, this might mean learning a new hobby or skill, such as a new language—perhaps even alongside your children!

Time for Reflection

Let's make our days more active together!

Let's Get Moving

- Recall active times from yesterday: dancing, playing, walking?
- Were there long sitting moments? Like watching a film?

Active Adventures

- Think of three activities everyone enjoys.
- Set a simple goal: "Skip or hop daily" or "Dance to a song before bed".
- Quick game: Who can tidy up their toys the quickest?

Even simpler still, plan a fun activity or meal for the coming week. Maybe it's "Taste a new vegetable on Tuesday" or "Have a family dance session on Saturday morning". The idea is to enjoy the journey to healthier choices as a family!

Walking through life side by side with our kids is like learning a dance—one where we lead at times and follow at others. The food we eat, how we deal with tough times, and our daily habits all play a role in helping us be the best parents we can and helping our kids grow up strong and happy. Moving around and always trying to learn helps us feel good and sets a great example for our kids. While things like staying active are super important for a healthy brain, there's so much more to the story. One big thing we often overlook: a good night's sleep. Sleep affects our energy, our food choices, how we handle stress, and even how we learn new things. In the next chapter, we'll dive deeper into the world of sleep and discover why it might be the missing piece to feeling our best. Let's journey together into the secrets of a restful night.

Chapter 17

BEING SLEEP WISE

Sweet Dreams

In Brief

Sleep is the number one thing anyone can do to improve their brain health and overall health. And for parents and caregivers, prioritising healthy sleep habits is essential for raising healthy children. This includes establishing consistent sleep routines, creating a calming sleep environment, limiting screen time before bed, and ensuring that our children are getting enough sleep for their age and development stage.

But if you found yourself chuckling or raising an eyebrow in sheer disbelief at these claims, you're not alone. To the uninitiated, it may sound like a grand overstatement. To those in the trenches of parenthood, it's a daily truth. Many of us can attest to the fact that when we (or our children) lack sleep, we don't just feel grumpy; our emotions teeter precariously on the edge. Our patience wears thin. We snap at the drop of a hat. And our rational thinking may as well be off on vacation. Sound familiar?

Welcome to the world of sleep-deprived parenting. But don't fret! By the end of this chapter, we will turn the tables on this exhausting foe.

Think of sleep not as a luxury but as the ultimate parental fuel. Just as a car can't run without gas, parents can't thrive without a good night's sleep. We're not just talking about surviving on a few hours of shut-eye here and there. No; we're talking about high-quality, restful, uninterrupted sleep that leaves you refreshed and ready to tackle the next day's joys and challenges.

For our children, sleep is even more imperative for the optimal development of their brains and neural networks. Children who get an adequate amount of sleep regularly experience improved attention, learning, and memory capabilities. For the developing brain, sleep is non-negotiable (despite how much our children might try to convince us otherwise!).

Now, the hard part is *getting* that sleep. And the hard truth is that during the first few years of your child's life, this may be an impossibility for most parents. Necessary wakeups for feeding or a horrifying nightmare can leave our children crying for us in the middle of the night or running to our beds, interfering with our own sleep. Unfortunately, there's no real way around these situations. We simply must do the best we can, such as heading to bed before complete and utter exhaustion sets in and trying to maintain our composure the best we can throughout our days, such as by being less reactive and more proactive.

At the end of the day, thriving parents understand the importance of sleep and make it a priority, not an afterthought. We encourage regular bedtimes for our children and ourselves. We strive to put sleep at the top of our list. So, let's try to understand the importance of this at a deeper level. Below, you'll find the ins and outs of sleep, as well as simple ways to pave your and your children's way towards a peaceful slumber.

Resting and Resetting the Brain: What Is Sleep?

As more and more research studies come out, we're learning sleep is non-negotiable for us and our children. But for most, sleep is a profound mystery. In fact, I can recall various chaotic and busy times in my own life when sleep was the first thing to go.

We might push through for the sake of work or schooling, thinking we are doing a world of good by "getting things done". But here's the kicker: when we cut back on sleep, we become less effective and efficient. Our cognition takes a serious hit, much more than most of us realise.

So, what's happening when we sleep?

In many ways, sleep is an altered state of consciousness. In the simplest of explanations, sleep is the housekeeper that comes and does a good wipe down of every nook and cranny in the home. But like any good housekeeper, sleep needs a few tools to do its job right, such as getting to bed at a decent hour and sleeping in a cool, dark, and quiet room.

During sleep, our body goes through various cycles, including Rapid Eye Movement (REM) and deep sleep. While we eventually do cycle through all the stages of sleep, the first half of our night is dominated by deep sleep, and the second half of our night is primarily REM sleep.

During deep sleep, the growth hormone is released, and our body focuses on tissue repair or development (in the case of our children). Our muscles relax, and there's an increase in blood supply to them. Our heart rate and breathing rate also slow down. If we don't get enough deep sleep, we are more likely to feel physically tired, and most of us need about 1.5–2 hours of deep sleep per night out of the recommended 7–8 hours of total sleep. In fact, not enough deep sleep, over time, can stunt our children's growth.

REM sleep, in contrast, activates some parts of the brain 30% more than when you are awake. In fact, brain-wave recordings comparing wakefulness and REM sleep are quite hard to differentiate between due to this heightened activity. As we enter REM sleep, the brain stem sends a signal down the spinal cord that paralyses voluntary muscle movements. This allows the brain to dream safely while it processes emotional memories. This is important for learning, attention, and memory. Without it, our children may become more impulsive, act out, be more aggressive, and have other behavioural and learning difficulties.

How Sleep Shapes the Developing Brain

Perhaps unsurprisingly, early sleep patterns, or rest and wakefulness states, develop in the womb. Early on, usually in the first six months of life, sleep stages are categorised as active sleep, quiet sleep, and indeterminate sleep. Eventually, after six months, sleep transforms into REM and deep-sleep cycles that are more akin to those in adults.

From the newborn stage to preschool age, the amount of sleep our children need dramatically changes. They may forego their nap, meaning it's no longer needed, and they begin to take on the sleep-wake cycle like adults: they are awake during the daytime hours and asleep at night.

While genetics plays a role in determining the quality and length of our sleep, those first few years (and months) play a major part in the development of the circadian rhythm; babies aren't born with this! And this is why many parents, if they have the chance, may recruit sleep therapists to help them determine a proper sleep schedule for their child when all else fails. These first few years are crucial for setting them up to have ample sleep and rest across their lifetime.

As you might already know, a consistent sleep schedule, a safe, comfortable, dark, quiet, and cool sleep environment, avoidance of digital screens, reducing stress, and regular daily activity all play a significant role in setting us up for sleep success. Since this information is attainable almost anywhere and is common knowledge, we won't go on about these factors too much, but it's worth noting their undeniable importance.

For the new parent, it might, however, be worth noting that allowing our children to fall asleep when still drowsy (i.e. putting them down in their cot before they fall asleep) helps them fall asleep on their own without the need for parental intervention, such as rocking them to sleep. This fosters healthy sleep association and can make bedtime less of a negotiation.

So, we've laid the groundwork. Now, how exactly does sleep impact the developing brain?

Sleep protects our memory and learning skills, *and* helps develop them. Children who obtain good quality, restful sleep tend to have better working memory, making learning easier. In contrast, sleep deprivation is linked to poorer academic performance. Additionally, naps during the infant years help with long-term memory, which is vital for learning and retaining language and other skills. Researchers hypothesise that naps are essential for infants for this exact reason. Children who sleep more restfully at night also tend to learn language skills easier and faster.

Sleep is not only important for memory and learning but also for creative-thinking abilities. More and more studies come out all the time on these aspects, and many argue that the definition of "creative thinking" is vague and tough to pinpoint. However, those who get enough quality sleep seem to show more creative-thinking abilities, such as better problem-solving skills.

As we all know, sleep also plays an important role in emotional regulation. As adults, we know all too well the irritation and aggravation we feel when we haven't had enough sleep. We do things we may regret; we say things we don't mean (sound familiar?). More and more studies indicate that less sleep leads to more negative feelings while more sleep leads to more positive emotions. With REM sleep accounting for a large portion of children's sleep, it's thought that this part of sleep is a major facilitator when it comes to emotional regulation and learning. In fact, a lack of sleep may increase activity in the amygdala, the fear centre of our brain, according to some studies—suggesting the root cause of this emotional reactivity we all feel when we lack sleep is that the smoke detector is setting off more false alarms.

Sleep disturbances might affect brain structure. Surprisingly, few studies examine the association between the developing brain and sleep. But this isn't to say that this association is not important. Out of the small amount of research that has been done, sleep disturbances during the first few years of life were shown to be associated with structural changes, such as reduced grey matter and a thinner cortex in the dorsolateral prefrontal area (an area of the brain associated with attention and learning). In turn, we can draw some conclusions and safely say that a lack of sleep has negative effects on the developing brain and its proper maturation.

Sleep issues are also implicated in the eventual development of mental health disorders like depression, anxiety, and more. In fact, sleep problems and mental health issues in childhood and adolescence tend to occur together, even becoming a vicious cycle where one impacts the other and vice versa.

Parents and Sleep

At some point in time, most parents find themselves sleep deprived. This, unfortunately, is a given, and there's usually no real way around it besides pushing through, knowing this time is temporary, and simply doing the best we can. We don't need to feel guilty about this.

At the same time, we can find some relief by trying to go to bed early and perhaps napping when possible (and *if* possible).

However, once your children reach an age where they are sleeping through the night, we can get back on track with our own sleep and use this as one of our biggest tools as parents. When we get adequate sleep, we become proactive as opposed to reactive. We have more motivation to ensure our family is eating healthily and moving their bodies regularly. In many ways, all aspects of health intertwine around our ability to sleep well.

And here's another twist: studies show our sleep patterns impact our children, and our children's sleep patterns impact us. It's a two-way street. Yet, as parents, we can lead by example, just as we can with dietary and exercise habits. We can set the stage by limiting social media and digital device usage around bedtime. We can foster healthy sleep habits by demonstrating and involving our children in healthy bedtime routines. And we can build our bedtime habits with our children based on the serve and return relationship.

How to Become Sleep Savvy

Here are some tried-and-true strategies that have helped many parents reclaim their nights, refuel their days, and foster healthy sleep habits for their children.

The "Simple" Pieces

As mentioned above, nothing quite beats a set sleep schedule, a restful environment, a good diet, regular exercise, and boundaries regarding digital devices close to bedtime.

I'll make this brief: your body loves routine. If we go to bed and wake up at the same time each day, the routine gets easier and easier as time goes on. Regularity provides an anchor for sleep, improving both quality and quantity.

Creating a restful environment, such as keeping your bedroom dark, quiet, and cool, can also work wonders, despite how small such factors may sound. Investing in a good mattress, pillows, and bedding also plays a role here, as does ensuring your bed is used only for sleep. Avoid working in bed or doing other activities that aren't associated with sleep in this area.

While we already have a whole chapter dedicated to technology, a reminder never hurts; turn off your electronic devices or create boundaries around them. Set a rule that no one can use them in the two hours leading up to bedtime. Instead, engage in some of the activities as a family that are outlined below.

Lastly, your diet plays a big role too. We want to avoid heavy meals, alcohol, and caffeine close to bedtime. While we might think a night-cap will help us sleep, this is far from the truth. Alcohol acts like a sedative. It knocks us out but doesn't boost quality sleep, which is why we feel like we slept well but also feel exhausted and cranky. Put simply, alcohol wrecks REM sleep, hindering our learning, memory, and emotional regulation abilities.

The Lesser-Known Sleep Tips for Healthier and Happier Children

Now that we've got the basics out of the way, let's turn to the lesser-known things about becoming sleep savvy. As we already know, the serve and return relationship is crucial for our children's development. This relationship helps them feel seen early on, guiding them towards becoming healthy, happy, and successful adults.

And we can nurture this type of relationship as part of our bedtime ritual. We can read together, play a relaxing game together, or do any other winding-down activity while ensuring we give our children our full love and attention. Many times, this doesn't have to be prolonged; even 15 or 30 minutes before their bedtime works. As parents, if we can always come back to keeping the serve and return relationship at the forefront, we can help our children avoid mental challenges later in life and truly thrive. Love is truly the antidote for almost everything, and showing your child unconditional love, affection, connection, and attention before bed can help them feel relaxed and help guide them towards a better sleep.

If you've read this far, congratulations! You're already on your way to becoming a sleep-savvy parent. The journey may be challenging, but remember, every small step you take towards better sleep is a step towards thriving parenthood and a happy and healthy child.

As you close this chapter, keep in mind that there are few constants, but the importance of rest—for both you and your child—is undeniable. Remember, recharging isn't a luxury; it's a pillar of well-being. So, as you lie down tonight, know that you've done your best. You've braved the challenges and celebrated the joys. Sleep with the assurance that you are paving a brighter, healthier future for your child. Sweet dreams, and well done on another day beautifully navigated.

Time for Reflection

This exercise is designed to help you reflect on your current sleep habits and make conscious decisions to enhance your night's rest. A good night's sleep plays a crucial role in our physical and mental well-being.

Rituals and Routines

- Reflect on your current bedtime routine. Do you have a set time for sleep and waking up?
- How often do you stick to this schedule during weekdays? Weekends?

Sleep Environment

- Look at your bedroom. Does it promote sleep? Is it dark, quiet, and cool?

- Think about your mattress and pillows. Are they comfortable? How old are they?
- What activities do you usually do in bed? Reading? Watching TV? Working?

Tech Boundaries

- Recall the last three nights. How often did you use electronic devices before bed?
- Challenge: can you set a goal to reduce screen time an hour before sleep? What other relaxing activities can replace it?

Dietary Choices and Sleep

- Reflect on your diet in the evening. Do you have heavy meals, alcohol, or caffeine close to bedtime?
- How do you feel in the morning after such meals or drinks? Refreshed or groggy?

Active Mind

- When you're in bed, do thoughts or worries keep you awake?
- Consider introducing relaxing practices like reading, meditation, or soft music to help calm the mind before sleep.

After you've thought about the above points, jot down three simple changes you'd like to introduce this week. Maybe it's setting a consistent bedtime, investing in blackout curtains, or trying a new bedtime ritual such as reading. Remember, every little change can bring you one step closer to restful nights and energised days.

Chapter 18

PRIORITISE YOUR BRAIN HEALTH

Never Lose Sight of What Truly Matters

In Brief

When we feel healthy and well, it positively influences every facet of our lives, including parenting. Being in a sound mental state equips us with the patience, understanding, and energy required to care for and nurture our children effectively. Healthy parents often raise healthier children—both mentally and physically.

Parenting is a demanding role, filled with its unique set of challenges. When we prioritise our brain health, we are better equipped to handle these challenges—from managing the daily routines to making critical decisions for our children's future. Moreover, children often mirror the behaviour and attitudes of their parents. By maintaining and showcasing a focus on brain health, parents can model positive behaviours and attitudes towards mental well-being for their children. This not only sets the foundation for their children's future mental health but also creates an environment where discussing and prioritising brain health becomes a norm.

Brain health is not just about preventing disorders; it's about enhancing our capacity to think, learn, and engage with the world around us. As parents, when we are mentally fit, we can provide our children with richer experiences, more meaningful interactions, and a more stable and loving environment. Making brain health everyone's business starts with personal responsibility. It's not just about the broader societal implications; it's about the immediate positive ripple effects it has on our families, especially our children. By understanding and emphasising the importance of mental well-being, we can be better parents today and lay the groundwork for the healthier generations of tomorrow.

Brain Health Is Everyone's Business

Imagine looking at yourself, your family and your children through a different lens—one that focuses on brain health as a crucial aspect of overall well-being. Let's move the dial on mental health disorders, prevention, and treatment; let's start to think about the brain and *talk* about the steps we are taking to support brain health in our family, children and community. Start small by adding it to conversations about going to the gym or on holidays. Have you wondered why we don't discuss brain health like we talk about heart health or physical health? And yet brain health is at the epicentre of our overall health.

According to WHO, brain health allows an individual to realise their full potential throughout their life. This holds true regardless of whether they have any specific disorders. When we focus on enhancing our brain health and function through exercises and activities, the mindset naturally follows. This is why looking at brain health and development first is a much better strategy than trying to strong-arm our mindset by doing things that don't necessarily improve it. When we train our brains, especially in terms of stress, we alter our mindset—without a single thought towards it. This chapter aims to empower you with the science and practical tools you need to make brain health a priority for everyone in your family. Our hope is that this new perspective leaves you feeling that both you and children are being seen and nurtured in a way that actively promotes, rather than hinders, your mental health and well-being.

Chapter 18: PRIORITISE YOUR BRAIN HEALTH

The hope is that moving the conversation to one around brain health and fitness may go towards destigmatising mental illness for everyone. Promoting the brain's untapped capacity for mental strength and resilience may help build our capacity to flatten the mental health crisis. Various factors, ranging from physical health and safe environments to lifelong learning opportunities and social connections, shape how our brains grow, adapt, and respond to stress. By proactively addressing these determinants, we can not only optimise brain health but also foster positive social and economic impacts that contribute to the well-being of society.

However, it's important to recognise that conditions affecting brain and nervous system health can manifest at any stage of life. These conditions often result in disrupted brain development, structural damage, or impaired functionality. Such conditions could include congenital issues, neurodevelopmental disorders, or neurological disorders that might arise at any point in one's life. Managing these conditions necessitates a multisectoral, interdisciplinary approach from a full spectrum of healthcare professionals and carers—focusing on promotion, prevention, treatment, and rehabilitation. Equally important is the active involvement of individuals with lived experience, along with their families and caregivers, in shaping their own holistic, person-centred care plans.

How we grow and thrive as humans means facing an uncomfortable truth: adverse experiences in childhood aren't just isolated incidents. They can be passed on from one person to another, almost in the way a cold is transmitted. Here's a simple comparison. Doctors often talk about stopping the spread of viruses from a parent or caregiver to their baby; we need to think the same way about stopping the spread of these childhood experiences. Our interactions with kids aren't just "in the moment"; they trigger real, physical changes in the brain and body that can affect them for the rest of their lives. It's not just what we say, but also how we look, smile, or even smell that can influence a child's development.

Understanding that our actions trigger a physical response in each other changes how we think about our own growth and well-being. It makes us reconsider how even small actions can have a big impact on someone else's life. By recognising this, we can focus not just on treating problems after they happen but on preventing them in the first place, making life better for everyone, now and in the future.

For the first time in history, we can see inside the living human brain, and it has revealed itself in shiny bright colours and billions of patterns. We can see something in a new way. Being seen through one's family history is not a hot topic for a neuroscientist to discover after decades of searching for answers about the causes of mental health disorders. It turns out it is not anyone's hot topic. We prefer to avoid being seen in this way and instead focus efforts on diet, exercise, relationships, sleep, all types of therapy, massage, cold showers, breathing exercises, psychedelics, silent retreats, religion, drugs, work, money, and doing hard things. Of course, these efforts are not wasted and are foundational to health and fitness.

We are not born a blank slate and do not have an equal starting place in life. There are many factors outside our control, such as whether we are born into wealth or poverty, with loving or hateful parents—or no parents—or grow up exposed to drug addiction, polluted or green communities, or have rare cancer mutations or long-living genes ... the list goes on.

These factors predetermine our start in life because of how these factors shape our nervous system. What was wired into the brain and body over centuries is not wired out overnight or through short-term solutions. The brain must be seen, discussed, and taught in a new way. The brain is the most important and least discussed part of the human body. To move forward as a society, we need to shift the focus across all sectors to prevention rather than only treatment and cure. Otherwise, we will unintentionally continue to pass on to the next generation the experiences of our fathers and mothers. This happens

Chapter 18: PRIORITISE YOUR BRAIN HEALTH

because the brain operates in primarily a default passive mode, without a remote control or manual. Imagine knowing you can train your brain like a muscle—or better still, that you can become the boss of your brain.

There is an emerging common language settling upon brain health with a focus on neuroplasticity. Science and exceptional people are demonstrating en masse that the brain can be directed and trained like a muscle to move forward from the past stress and trauma. Imagine being seen by a health practitioner, knowing you will leave with a personalised brain health and neuroplasticity action plan. Imagine that the plan is based on an understanding of you and your life and considers the impact of adverse childhood and adult experiences, diet, exercise, sleep patterns, the type of work you do, the number of social connections and supports you have, and your beliefs about sickness and healing.

And imagine that before writing the prescription for the right medication, the practitioner prepares a plan that constructs the steps and resources that leverage neuroplasticity. We know that the time allocated by health practitioners to each patient in general practice is 5–15 minutes, and of this they can listen to the patients for about 60 seconds; in counselling rooms it's about 45–60 minutes. Many health practitioners will tell you that people want a pill and not an action plan, and for their problem to be solved immediately rather than over time. The real aim for all of us as a society is to build preventative brain health plans that are part of a public health initiative to reduce the need to wait for and be seen after a crisis. This is the future that everyone is searching and hoping for. This is what *being seen* can become.

Given that a variety of chronic diseases are associated with a history of childhood maltreatment, it becomes increasingly clear that screening for such experiences should be a standard practice in adult medical care. Brain health isn't a sideline activity or someone else's problem—it's everyone's responsibility, like many other public health campaigns such as anti-smoking, sunscreen and mandatory seatbelt-wearing initiatives. This is of utmost priority; it will save countries and healthcare systems trillions of dollars. Preventing child maltreatment isn't just about stopping bad things from happening to our kids. It's about building a healthier, stronger society for everyone. Addressing the root causes—like family conflict, parental mental illness, and substance abuse—can make a monumental difference. Not just in preventing maltreatment, but also in reducing risky behaviours and mental illnesses that often follow. And let's zoom out for a moment. We should advocate for *social policies* that address those root causes of family stress, such as financial stress, poverty, homelessness, and unemployment. These are not just societal issues; they're health issues. This is a societal problem and requires preventative measures. We're not just doing social good for *now*; we're actively shaping life for future generations.

Looking back, if I had known in my youth what we know now about brain science, ACEs, and the brain's ability to change, my sister's life could have been different. But the good news is that we have the power to shape the future through neuroscientific advances, education, and changing the narrative around mental health. Sometimes, even a simple smile or a kind gesture can change someone's life direction.

Despite this compelling evidence that early life experiences contribute significantly to mental health disorders, the medical and allied health communities and society have been slow to fully recognise how these early life traumas can contribute to adult diseases. By understanding ourselves and our past experiences better, we can stop the cycle. This new awareness can help not just us but also future generations to reduce the impact of bad childhood experiences and mental health issues.

Going forward, it will be crucial to expand our understanding through more research, refine treatment methods, and develop effective social interventions that consider the impact and how to prevent or limit early life trauma. Public policies must also evolve to address this significant health concern.

By asking questions like, "Can you tell me about your childhood?" or "Were there significant events when you were young that you think have shaped you?", healthcare providers can help you uncover and articulate ACEs. This not only aids in diagnosis and treatment but also empowers patients to integrate this understanding into their own life narrative, thereby facilitating healing and growth. This borrows from the practice of "clinical yarning", which is a term used within Aboriginal communities in Australia to describe a form of storytelling that has therapeutic benefits. It offers valuable lessons for transforming healthcare. By embracing the power of stories, healthcare providers can forge deeper connections with patients, enhance communication, foster empathy, and discuss the steps to make significant change, rather than simply escalating medicines that fail in the end.

So, let's shine a light on this topic and see ourselves and others in a new, more understanding way. It's our best shot at making a meaningful difference for everyone's mental health. Instead, a more holistic approach that considers the role of ACEs or trauma is essential. Being seen through our family history can create a new future through healing ACEs. Imagine a world where every individual, irrespective of background or circumstance, has access to the mental health resources and support they need. Let's carry this hope forward.

Therefore, the responsibility for creating a paradigm shift towards health and resilience lies in each of us—scientists, parents, educators, and policymakers alike. The tools are within our grasp, from cutting-edge neuroscience to globally connected digital platforms. Together, they offer unprecedented opportunities to turn the dream of a fresh start into a lived reality for countless individuals. The promise of neuroplasticity offers hope for those recovering from ACEs, and it also establishes the scientific groundwork for pre-emptive measures to prevent these adverse experiences from echoing through subsequent generations. Now, more than ever, the opportunity to enact transformative changes is within reach, not as a distant aspiration but as an immediate, collective endeavour. Let us seize this moment to forge a path that not only heals but also perpetuates cycles of positive transformation that will reverberate through the generations to come.

The dawn of a new era in mental health awareness and treatment is on the horizon. As we move towards it, let us champion the cause, ensuring that these transformative breakthroughs find their rightful place in the limelight, benefiting humanity at large. This isn't just about our well-being today but about safeguarding the mental health of our children and grandchildren. Can there be any mission more vital? As we advocate for and adopt practices that foster brain health and mental well-being, we pave the way for a brighter future. Together, we are making brain health everyone's business and not someone else's problem. This proactive stance ensures that we are addressing immediate concerns *and* future-proofing our families. In doing so, we become change agents, paving the way for a brighter, healthier future for all generations. Yes, the dawn of a new era in mental health is upon us, and together, we can ensure that it heralds hope and well-being for all.

Conclusion

Imagine a world where your child looks up to you, not just to ask for the wi-fi password, but to seek wisdom, comfort, and connection. A world where technology serves as a bridge, not a barrier, between you and your child's inner universe. This isn't a utopian fantasy; it's a reachable reality, and this book has been your roadmap. While we cannot turn back the clock to simpler times, we can adapt and evolve. We can update our parenting toolkits to include digital literacy, teaching ourselves and our children how to use technology as a tool, rather than accepting it as a tyrant. I hope this book equipped you with useful skills by blending insights from neuroscience with practical advice, providing you with actionable steps that prioritise brain health and overall well-being over mere screen time.

If you've made it this far, you're not just a parent; you're a parent on a mission. The mission being to reclaim the essence of parenting from the clutches of an ever-distracting digital world. Remember the story I shared in the introduction about the parent who felt she had lost her child to screens? She found her way back, and so can you. Parenting in the digital age is like navigating a maze, but what if I told you that you've been equipped with a compass? It's called "Being Seen", and it's not just your compass; it's your North Star.

In our journey through this book, we've explored how technology, an undeniable force in our lives, is both an asset and an obstacle. We've delved into the essential role of early life experiences in shaping your child's brain, affecting their mental and physical health for years to come. We've confronted the unsettling truth that screens and social media platforms are increasingly becoming the "eyes" through which our children are seen, often with implications we can't fully grasp. As digital-age parents, we must now extend this protective gaze to the online environments our children frequent, becoming adept at identifying signs of online exploitation and grooming. The task is daunting, but the stakes couldn't be higher.

Understanding the vital importance of "Being Seen"—of building strong, meaningful serve and return relationships with our children—can serve as a robust defence against the dangers lurking in the digital shadows. Such relationships not only forge stronger emotional bonds but also contribute to healthy brain development. Thus, while parenting in the digital age may be the hardest job we'll ever have, it also offers an unparalleled opportunity. By taking the time to genuinely see our children, we can offer them the most lasting and meaningful gift: a solid foundation for a life of well-being. A parent's undivided attention equips children with a sense of individuality, autonomy, and security, laying the foundation for a healthy brain and, by extension, a thriving adult life. Amid the pressures of modern life, the simplicity of this solution—giving our children the gift of being truly seen—is both attainable and profoundly impactful.

This book has handed you a new vocabulary, a digital language that you need to connect with your child. We've learned how to initiate open-ended questions and conversations that not only make your child think but also feel understood and valued. We've also tackled the awkward, challenging topics like technology and sex, understanding that these conversations are both inevitable and crucial. From being overwhelmed by the digital age, you can become proficient navigators of it. Your new skill set doesn't just include managing screen time or understanding social media algorithms; it involves the deeply human art of truly "seeing" your child and ensuring they feel seen by you. This is how you build a resilient, thriving child in a technology-dominated world.

So, why does this matter? It matters because every moment you invest in truly seeing your child is a stitch in the fabric of their future well-being. We are the architects of the next generation, and the tools are in our hands. However, the clock is ticking. Our children are growing up fast, and the digital world is evolving even faster. We cannot afford to waste time lamenting the challenges. Instead, let's use our newfound knowledge and skills to turn these challenges into opportunities.

Have you ever wondered why one person in a family might struggle with feelings of sadness or anxiety, while their sibling seems perfectly fine, even though they grew up together? It's a puzzle many of us have thought about. The tapestry of our lives is woven from the threads of our genetic makeup, the environment we are nurtured in, and the unique experiences we face. While my sister and I shared many similarities, our paths diverged in significant ways. Many ask why it was she who ended up in the hospital and not me. The answer is complex, and someday I'll share the complete story. But at its core, the difference can be attributed to the feeling of being truly seen, acknowledged, and enveloped in love.

When I was an infant, my grandmother played a pivotal role in my life. I was nurtured by her gentle rocking, the lullabies she sang, the warmth of her embrace. It was she who gave me the nickname "Seen". Though it might seem simple, to me, it was a testament to her ability to truly see who I was, to recognise the spirit within. Over the years, few have called me by that name, but each time I hear it, I'm transported back to those early days of unconditional love. But my grandmother wasn't the sole beacon of love in my life. I was fortunate to be surrounded by a village of support. Every hug, every word of encouragement, and every gesture of kindness fortified me against life's challenges. I would wish the same for you and your children and the generations to come.

As the chapters in this book unfolded, you weren't just reading; you were evolving. Whether it was understanding your own behaviour patterns or learning how to safeguard your child online, with each chapter you have found a stepping stone to becoming the parent you've always aspired to be. You've solved the problem of feeling lost in the digital age by equipping yourself with the skills and knowledge to not just survive but thrive in it. So, here we are, at the end of this book but at the beginning of a new chapter in your parenting journey. It's time to start co-authoring the next chapter, both in your life and in your child's.

As we close this book, let's open a new one with renewed hope and commitment where our children are being seen by us, loved by us, and prepared for whatever comes next, screens and all. The responsibility is immense, but so is the reward: the lifelong well-being, security, and happiness of our children.

As I type these final words, my thoughts drift to Francesca. Francesca never knew a world where "Being Seen" was understood and embraced. Today, as I see the big steps we're making, I'm filled with hope for a better future. Thank you for giving your children the important gift of "Being Seen". Your love, time, and attention are making the world a better place for everyone.

Acknowledgements

I would like to extend my heartfelt gratitude to many people who have played a pivotal role in the creation of this book. Your support, expertise, and unwavering commitment have been invaluable, and I am deeply thankful for your contributions.

First and foremost, I want to express my sincere appreciation to all the guests who have given their time to the *Thriving Minds* podcast. Your insights and knowledge have provided a deep well of knowledge that have played a role in my parenting journey and as a neuroscientist.

To the brilliant scientists, experts, professionals, parents, and educators who have shared their wisdom and experiences. Your collective efforts have enriched the content of this book, providing readers with a deeper understanding of the mind and its potential for transformation. I extend my gratitude to the brilliant minds whose technological innovations have revolutionised our understanding of the brain. Their work in computer technology, neuroimaging, genomics, data analysis, and now generative AI, has propelled our knowledge forward at a pace so rapid that textbooks struggle to keep up.

I extend a special thank you to Krista Bugden and Jane Smith for their expert assistance in the writing and copyediting process. Your meticulous attention to detail and guidance have been instrumental in shaping the clarity and quality of this work.

I am also indebted to the early readers of this book, who offered their valuable feedback and insights Hailey Bartholomew, Martin Betts, John Bray, Rebecca Foley, Lee Forbes, Sam Jockel, Rhiannon Phillips, Hannah Reed, Aaron Smith, Jackie Taylor who have refined the ideas ensuring that this book resonates with its audience.

Lastly, I want to acknowledge my husband Martin Betts and our five children, their partners, our extended families and friends. Your unwavering support, patience, and understanding during the process of writing and researching this book have meant the world to me. Thank you for being a constant source of motivation.

About the Author

Professor Selena Bartlett is a distinguished leader in brain science and leads the Neuroscience and Neuroplasticity Group at the Translational Research Institute, School of Clinical Sciences, Faculty of Health, Queensland University of Technology in Brisbane, Australia. Selena's contributions to neuroscience—acknowledged by the Lawrie Austin Award and three Women in Technology Awards—are complemented by her most rewarding role: parenthood.

With over 110 research papers, books, and numerous media appearances, Selena excels in translating complex science into practical insights. She is also qualified as a counsellor, pharmacist and a mathematician, having gained degrees in Pharmacy and Applied Science in Mathematics and a diploma of counselling. As host of the *Thriving Minds* podcast, she offers empowering advice and inspiring stories, making scientific wisdom accessible to all.

In this book, Selena shares her knowledge gained from thirty years of studying the brain, early life experiences and neuroplasticity. She presents a revolutionary blueprint that pivots from mental health to brain health—showing us that the brain, too, can grow, adapt, and heal. She combines the findings from evidence-based research with real-life examples, wisdom, and simple tools and tips that even the busiest of parents can adopt to help their children become happier, healthier, and stronger—at any stage in life.

www.ingramcontent.com/pod-product-compliance
Lightning Source LLC
Chambersburg PA
CBHW051605010526
44119CB00056B/786